THE EXTINCT

THE EXTINCTION CURVE

Growth and Globalisation in the Climate Endgame

BY

JOHN VAN DER VELDEN

ROB WHITE

emerald
PUBLISHING

United Kingdom – North America – Japan
India – Malaysia – China

Emerald Publishing Limited
Howard House, Wagon Lane, Bingley BD16 1WA, UK

First edition 2021

Reprints and permissions service
Contact: permissions@emeraldinsight.com

British Library Cataloguing in Publication Data
A catalogue record for this book is available from the British Library

ISBN: 978-1-80043-827-9 (Print)
ISBN: 978-1-80043-824-8 (Online)
ISBN: 978-1-83982-670-2 (Epub)

ISOQAR certified
Management System,
awarded to Emerald
for adherence to
Environmental
standard
ISO 14001:2004.

Certificate Number 1985
ISO 14001

INVESTOR IN PEOPLE

CONTENTS

ABOUT THE AUTHORS

John van der Velden is an independent socialist writer living in Canberra, Australia. He was a national convenor, on behalf of the Non-Aligned Caucus, in the initial multi-party Socialist Alliance. He writes on matters of political economy, class structure and the climate emergency.

Rob White is Distinguished Professor of Criminology at the University of Tasmania, Australia. Considered a pioneer in the field of green criminology, he has particular interest in transnational environmental crime and eco-justice. Related books include *Crimes Against Nature* (2008), *Transnational Environmental Crime* (2011), *Environmental Harm* (2013) and *Climate Change Criminology* (2018).

ACKNOWLEDGEMENTS

We acknowledge the Aboriginal and Torres Strait Islander peoples as the true custodians of the lands on which we live and work. The sovereignty of the more than 250 First Nations in 1788 has never been ceded. European invasion came with massive destruction of natural ecosystems, the widespread disruption of traditional caring for the country, and deliberate fracturing of deep spiritual connections between the First Peoples and their lands since the Creation Time. But the over 60,000 year organic cultural bonds with country are not severed and First Peoples resolve is unbowed. Demands for Treaty recognition of Indigenous sovereignty, custodianship, land rights and compensation on just terms remain unfulfilled. Their struggle against dispossession, colonisation and genocide and for self determination continues. It's an inspiration of resilience for us all.

We are grateful to a number of people for their advice, suggestions and feedback in the preparation of this book. Our sincere respect and thanks to Ngunnawal Elder Roslyn Brown for her guidance on the First Nations statement. Alastair Greig provided valuable early collaboration on material in chapter 2 and an insightful and considered assessment of the final manuscript. His input and friendship over the years is gratefully acknowledged. We are also particularly thankful to Lesley Hayes, who generously read the manuscript at multiple points and kept the drafting grounded in the political universe of the audience for which it is intended. Alyson and Michael Kakakios provided much appreciated and valued insights and encouragement, as did Stephen Richer, Janine Smith and Frances Medlock. Will Steffen provided guidance on the climate science, trends and issues for which we are likewise appreciative. The shortcomings are ours.

A big thanks as well goes to Jules Willan and the team at Emerald for their enthusiastic support for this project.

This book is dedicated to our grandchildren: Riley, Blake, Paige and Clayton (John) and Luca and Lola (Rob). We wish they were inheriting a better world than the one we stumbled into. Hopefully, they will find the wisdom and resolve to make a positive difference for their families and communities.

1

AT DANTE'S GATE

Plague is upon us. The global economy is in lockdown as yet another crisis in collective wellbeing unfolds. It is the second global economic crisis in as many decades. While this has started as a health crisis, it is following a familiar storyline. Governments are bailing out private companies by socialising their losses, while the social and economic losses of the global majority are being privatised.

The pandemic imagery of our class-divided lives is stark. In the advanced economies, those workers and shopkeepers more fortunate than frontline health workers are cocooned in relatively affluent bubbles of home delivery and debt quarantined enterprise. The super-rich are waiting it out on their yachts and holiday islands, bottom feeding on collapsing stock prices. Meanwhile minorities of colour are dying at more than double the national averages. In the emerging areas of the global economy the story is even more confronting. Workers leading day-to-day work-life are dropping dead in the streets. Social distancing and tips on health hygiene are gratuitous advice for those trapped in crowded slums without clean water.

In time, the health crisis will mostly pass. Survivors may have some degree of 'herd immunity' and a combination of treatment and immunisation will likely be put together for those less robust. Most importantly, in this most capitalist of storylines, the economy will come out of debt-fuelled Keynesian hibernation so that growth and accumulation can resume. Just don't mention the climate and ecological emergency. That replay button on further global catastrophes to come is a story sequel that our overseer capitalist editors won't publish. It remains an inconvenient truth.

This book addresses the capitalist growth and accumulation pandemic that lies at the core of our globalised crises. It is what has brought us to this calamitous historical point. Even with an immediate fundamental change in course, the next global catastrophe is already in the pipeline. But that fundamental change in course has yet to eventuate. Ominously, the climate and ecological *extinction curve* continues along its steep trajectory.

In this sense, given present upward directions, we are not offering a book of hope. But nor is it a requiem about our species and paradise lost. It is a sober assessment of the reasons we have entered a period of escalating crises in economic and ecological wellbeing on a global scale. These crises foreshadow a crossroads in the historical expansion and progress of our human civilisation. They reflect the consequences of an economic system grounded in maximum social exploitation of a finite natural world. The era of unsustainable growth has to come to an end. This book aims to explain why and how.

OUR HISTORICAL MOMENT

Historically, capitalism has been a source of breath-taking prosperity for some and a beacon of material aspiration for the global majority. This capitalist mode of production (CMP) is also a system of irreconcilable structural contradictions and crisis dynamics. These dynamics, and the swirling political vortexes they engender, increasingly constitute a sword of Damocles that imperils the continued social and political legitimacy of our national and international ruling elites. But it is also a sword that hangs over working people globally.

Economic and ecological conditions are destined to become more precarious for those outside the affluent zones of the global economy. Within the more buffered advanced economies, political and ideological consent is fracturing around what the future may hold with regard to structural shocks and dislocations, and how best to respond. Certainly, the consequences of capitalist crises of accumulation, globalisation and the climate/ecological emergency are weighing heavily on national and global political culture, even if the contours are not fully understood or accepted. On the longer-term horizon, the future welfare of vast sections of the global community is in question.

Catastrophic global heating, initiated by the capitalist industrial revolution, has tipped the earth outside the 12,000-year stable Holocene climate conditions within which civilisations have flourished. Projections for greenhouse gas and temperature accelerations are foreboding in their consequences for extensive habitable zones of the planet. Concurrently, capitalist globalisation has created staggering inequality, widespread species and ecosystem collapse, unfettered institutions of entrenched class power and interests, and an interdependent web of subordination to market forces beyond democratic control. It adds up to a perfect storm of biblical proportions.

Human civilisation inherited Holocene natural world conditions and blossomed materially and numerically. But, since the advent of the Capitalist Industrial Revolution in 1750, and subsequent globalisation of capitalism in particular, the inherited gifts of our natural world are on the brink of being trashed beyond recovery within any geological time frame that will matter to us and our descendants. Significant transformations are locked in and irreversible, though the pace and scale of impending catastrophe are not yet settled. Much is still able to be contested, politically and technologically. But adaptation, mitigation and survival imperatives will exponentially loom larger in the years ahead and stark choices will become more sharply defined. 'Business as usual' can no longer be accommodated. Neither can 'politics as usual'.

The future is still only uncertain for that hollow characterisation 'us as a species' – an expression which puts humans-in-general at the centre of problem, consequence and solution. This imprecise terminology is partly why we decline to use the latest 'Anthropocene Epoch' designation now commonly used in contemporary environmental discourse. The broad scientific intent is to describe the accelerated measurable technological impact of human activity on planetary eco-systems. The original intent was to particularly identify the human-driven trajectory of the Earth System out of the Holocene, of which climate science is an important part. However, it obscures if not misses entirely the capitalist origins, divergent class interests, and unequal consequences of that activity in generating climate and ecological catastrophe.

A principal intent of the use of the term Anthropocene by the climate action movement is to contest the politically motivated, self-interested denial of climate science. However, escalating ecological crises risk the ideological morphing of 'Anthropocene' from focusing on 'culpable human responsibility' and the absence of care for planetary eco-systems to the assertion of geo-engineered human ownership and control of nature. This shift is already evident. Geo-engineering 'solutions' to the climate crisis should be vigorously resisted, just for the cataclysmic risk levels alone. Moreover, they fail to address the physical causes of global heating, the need to restore eco-systems, and divert attention from the underlying political economy driving climate change.

The future is grim if not dire for vast numbers of us. The per capita carbon emissions in major developing countries are miniscule compared to that generated by the 1.5 billion people in the advanced economies. Just for the three billion in China and India alone, rudimentary expansion to 'first world' living standards is already swamping the remnant capacity of planetary carbon sinks to cope. Plus, there are still some three billion left in the remaining developing countries of the global economy whose fate is not being factored into this consideration.

Globalisation of capitalist growth has become a cascading set of wicked dilemmas that no government dares to fully confront. Entrenched economic power and dominant class interests are threatened by the questioning of capitalist growth. It likewise threatens the very institutional fabric of any existing individual nation state apparatus and elected government that are fundamentally compelled to serve those interests. Even many dedicated climate action advocates are reluctant to fully question or challenge capitalist growth for fear of where it might lead politically.

But confront these dilemmas we must. We urgently need to develop a clear navigational chart or there will be no political direction toward a future world any of us would care to embrace. The character of the contradictory forces and material self-interests driving this perilous climate endgame have never been more transparent or understood. Yet, paradoxically, despite this collective self-awareness, the widespread acceptance of paralysing obfuscations, diversions and inactions swirling around core climate problems have also never been more complete. Picking through this structural and political conundrum is an essential and urgent task.

US, HERE, NOW

There are confronting and stark political choices unfolding for our diverse green and broad left movements seeking to address this historical convergence. Despite decades of environmental and social action against the consequences of neo-liberal globalisation we still only have a nascent and fragmented movement for genuine climate action. The year 2020 marks 50 years of Earth Day. Yet in that time a consolidated and united challenge to escalating ecological catastrophe has failed to materialise. It follows from the faded political challenge stemming from the Global Financial Crisis (GFC) exposure of both the systemic fragilities of global capitalism, and the political bankruptcy and self-interest of transnational global elites. Ultra-Right conservatism is on the rise.

Now, as this current economic crisis assumes dimensions of depression to rival the 'Great One' of the 1930s, we still find ourselves politically behind rather than ahead of the extinction curve. Most progressive demands, actions and options still revolve around the question of whether a form of green capitalism is either possible or desirable. It is past time to find a common broad left course to settle this question.

This book hopes to contribute to that outcome. It has been in gestation for several years, framed by our acute self-awareness of generational time. Rob had just become a grandparent for the first time. Meanwhile, having flown into the brown haze covering continental Europe with partner, daughter and grandson, John found himself in the sculpture garden of the Rodin Museum. He had just read Jim Hansen's *Storms of My Grandchildren*. In the suffocating heat of a particularly hot Parisian July, standing in front of Dante's Gates of Hell gave pause to reflect. As we hurtle past the Holocene edge will our grandchildren bear witness to the end of our beginnings or the beginnings of our end? Whether with reflective hope or despair, our generations of the past 70 years have brought them to this gate. They are being shoved through regardless ... into the dark unknown. The political is very personal, and so is this book.

Collectively our progressive movement, like our class-divided global community as a whole, has arrived at a unique historical conjuncture. In generational terms we are at Dante's Gate and out of time to avoid passing through. What we do, or fail to do, in the next couple of decades will shape the immediate future for billions unlike any previous historical period. The stakes really are that high.

Our personal histories and collaboration parallel this historically unique generational convergence. We are part of the tail end of the post-World War II baby boom. Our parents' generation lived through the grim social and economic austerity of the European war experience, its preceding great economic depression of the 1930s, and then the explosion in capitalist production and consumption that followed post-war. Like so many working-class families of this era, our parents' experiences of depression and war framed their feelings of fortune and relief at the peace and prosperity which followed. But it was us Baby Boomer children that were the actual main beneficiaries.

Our working-class generation had historically unique affluence without the parental filter of genuine austerity. We benefitted from publicly funded infrastructure, education, health and social services, and high wages compared to the global norm. The productivity and competitive advantage of the advanced economies, which flowed through to the relative affluence of our generation and that of our children, is only now set to being reversed in the following generations.

Over the next few decades our generation of Boomer elders will be extinct, whether we rebel or not. For our grandchildren, like Greta Thunberg and the School Strikers for Climate, everything we have taken for granted is at stake. What we do, or fail to do, collectively will be decisive in shaping the roadmap for their lives.

In the midst of historically startling affluence within the advanced economies, three major additional forces have shaped our personal experiences and that of our generation.

Firstly, some 70 years of largely unfettered capitalist growth in the advanced economies coincided with a dramatic surge in ecological degradation and population expansion globally. Our parents grew up and established families in a pre-war world of 2 billion, up from 1 billion at the time of the capitalist industrial revolution in Europe in 1750. By 1960, this had jumped to 3 billion. During our personal life histories, this has exploded to 7 billion. The children and grandchildren in our families will share the planet with more than 9 billion on current projections. It could also be substantially less if the grim scenarios in this book are realised. Either way, the challenge of this integrated exponential impact of seemingly endless economic growth, ecological degradation, and population increase is epoch shaping.

Secondly, we have witnessed dramatic geo-political reconfigurations flowing from this rampant capitalist globalisation. Our personal social and political awareness was fashioned in the triangulated shadow of the reinvigorated capitalist imperialism of the American post-war hegemony versus the degenerated Stalinised 'Communism' of the Soviet Union (USSR) and Peoples Republic of China, which framed the decolonisation and 'national liberation' movements ending the 500-year European imperialist empire capitalist period. This international triangulation dominated our formative years until the demise of the USSR in the period 1989–1991. Post-1990, capitalist globalisation has embarked on a qualitatively new phase. Likewise, ecological degradation has been transformed qualitatively.

The seeds of contemporary capitalist globalisation were fashioned in the context of post-war American capitalism as the dominant and only capitalist superpower in relation to its war allies as well as the vanquished. The consolidation of the USSR, its Eastern European satellites, and the 1949 revolution in China established the post-war territorial limit of capitalist expansion under American capital domination. However, by virtue of their politically antagonistic relationship with 'Western' capitalism per se, the existence of 'communist' States also facilitated the increasingly borderless expansion, interpenetration and competitive integration of American finance capital across the balance of the capitalist world: Western Europe, Australia and Japan of course, but also the Americas generally, South Korea, Taiwan and other newly developing sectors of the capitalist world economy.

The result was expansion in depth and breadth of more globalised financing of commodity production, distribution, exchange and consumption. It also became the foundation for globalising transnational corporations and competitively integrated national and international economic elites.

While the relative post-war domination of American global capital has continued throughout this transnational globalisation it (ironically) started eroding at an enhanced pace with the demise and full restoration of capitalism in the USSR and its Stalinised Eastern European satellites (circa post-1989). Equally, if not more profoundly, the parallel restoration and global integration of Chinese capitalism consolidated and has now elevated contemporary forms of globalisation to a new highest expression of the capitalist mode of production (CMP).

The initial principal dominance of American capital has now been further diminished by multi-polar centres and largely unfettered corporate internationalisation of global capitalist economic power. The largest transnational corporations dwarf the financial resources and power of most countries and render 'national sovereignty' a misnomer for all but the largest nation-states. This explosion of capitalist growth has been accompanied by consequent growth of ecologically untenable, materially enhanced population levels, the explosion of greenhouse gas emissions, and widespread ecological degradation.

Thirdly, we have lived through and been active players in the shortcomings and failures to build effective fightbacks and alternatives to our present conjuncture of dilemmas. It's difficult to write off 50 years of collective (and personal) campaigning, marching, organising and writing as failure. Yes, there have been successes, even substantial ones. Strikes have been won, greater civil and women's rights achieved, dams have been stopped and wars have been ended. Still, we on 'The Left' as a whole have not achieved our transformational objectives – whether that objective has been fundamental reform to make global capitalism fairer and more equal, a revolution to overthrow capitalist power and exploitation altogether or stopping ecological destruction. If anything, the achievements of hard-won struggles are under robust challenge and the political risk levels threatening a century of progressive social reforms are escalating. 'The Left' remains in a long-term state of crisis and decline.

THE CLIMB

Over the past 40 years, there has been widespread social democratic centrist government capitulation (if not support) to expand capitalist neo-liberalism, globalisation and geo-politically strategic war. This has generated a crisis of 'progressive' political legitimacy for 'reformist' social democracy across the advanced economies. For example, it was a British Labour government which followed America into a disastrous military quagmire in Iraq. An Australian Labor government floated the Australian dollar, sold off public assets, lowered tariffs and internationalised the economy with subsequent re-location/demise of domestic manufacturing.

The capitalist managerialism of social democracy has been consistent and persistent across the liberal democracies. The confusion and corrosion of traditional support within the working class has followed. The Left inside many traditional social democratic parties has been largely marginalised, eroded or departed altogether. It reflects a wider pattern of disgust and mistrust of self-serving 'business as usual' globalised elites and their 'political class' across subordinate classes. The prospect of a successful left social democratic reform agenda succeeding in coalition with centrist social democracy is thereby a challenging leap of faith. It is beyond serious credibility unless the ultimate aim is to simply resign ourselves to greenwashing the economic system and structure of power that got us to this climate endgame.

Concurrently, Stalinisation of 'socialist type' revolutions, including those beyond its initial Russian borders (Eastern Europe, China) has deepened the long-running existential crisis within traditional Leninist-based revolutionary socialist left alternatives, particularly post-1989–1991 capitalist restorations. Growing support for Ultra-Right narratives against centrist elites among the working class in the advanced economies is circumventing traditional revolutionary socialist narratives. For traditional, anti-Stalinist revolutionary socialists it punctuates a gallant and noble but rather insular, compromised, ineffective and all too dismal legacy.

An eco-socialist future is by no means unattainable. We will be arguing to the contrary. It is both necessary and possible. More so even, democratic socialist reorganisation of the national and global economy is one of the few options available to respond to the seemingly intractable ecological problems before us. We will argue it is certainly the more attractive democratic option. But the prospect of a more immediate extreme rightward political response to the shrinking privileges of capitalist globalisation is ominous. New pathways to a democratic eco-socialist transformation are required.

There are no real do-overs, fresh starts or clean slates in politics. The emergence of a post-truth, 'fake news' political culture is unlikely to facilitate a successful historical re-write for the Stalinist heritage within the broad socialist left in particular. In large measure, we are where we are. There are learnings to be had and reflect upon, which we do below. On balance however, these political and ideological shortcomings and failings are not the decisive factor. Fundamentally, the subjective force of political struggle rarely rises beyond the objective conditions of what is possible. Just fulfilling what is *possible* is rare enough historically.

In that context, looking back and taking stock of the Left political landscape of the past 70 years, two things stand out. Firstly, our working-class base in the advanced economies is largely too affluent, its political struggles too easily blunted and bought by relative national prosperity and has too much 'advantage' to risk compared to the working class globally. Much as we green/socialist left have expected that to change in time it hasn't sufficiently happened yet for a transformational message to germinate and grow into a transformational force.

Secondly, by extension, much the same could be said of the limited active supporter base across the Left spectrum. The cultural hold of social democracy as an aspirational political alternative among even resilient left political supporters remains remarkable. With all due respect to the dedicated left social democratic reform wings across the advanced economies, social democracy is and always has been a form of *capitalist* governance. Getting voted in to run a capitalist economy is to be a capitalist government. If the transformational message isn't compromised getting there, then for sure the transformational action agenda will be post-arrival. It is a history repeated in diverse countries and circumstances.

Much as there had been real momentum (now in doubt) for a Green New Deal in the 2020 American election, and this should still be *critically* supported across the Left, these compromising trajectories need to be acknowledged. We have been down this incremental 'politically realistic' reform road before. It has largely shaped the broad Left agenda and historical experience over the last 100 years. We only have to briefly reflect upon the de-mobilisation and effective demise of the GFC-generated Occupy Movement under eight years of Obama and Democratic Party centrism to see the political warnings.

Macron, Trudeau, Rudd-Gillard, Obama, even Clinton, are clearly preferable governments to that offered by Le Pen, Abbott-Morrison, Johnson and Trump. But that would be to miss the point. The prospect of a kinder, more equitable, more ecologically sustainable and controllable capitalism via the ballot box retains a deep hold. When the 'partial victories' of social democracy do occur, they have rarely become the energiser for wider transformational political ambition for our green/socialist supporters or our diverse working-class base. Most often the opposite has been the case.

The surprise at Trump and Brexit is remarkable for its surprise. The large section of 'left behind' disaffected working-class *support* for both reflects a level of structural despair and disgust for 'politics as usual', well beyond that of our diverse left supporter base. Affluent Left political blindness is playing its part but so too is residual faith among our support base for the institutional integrity of 'bourgeois liberal democracy'. Trump's 'base' is cheering his ideological flouting and dismantling of the institutional and constitutional icons of the American capitalist state. Vast layers of the American Left, and globally, are hoping and cheering that these same (traditionally bête noire) institutions find a way to bring him down. Strange days indeed.

Capitalist globalisation has ensnared us all in an insidious web of endless growth and expanded consumption, accelerated carbon footprints and temperature extremes, and all-encompassing ecological degradation.

If there is to be hope it will be created by an increasingly disenfranchised global majority eventually being forced to democratically assert its (our) collective self-interest. This will only eventuate amid collapsing ecological and economic conditions that are not of our choosing.

Ecological dispossessions and economic calamities are not abstractions. They are real and in the here and now – in the scale of the health and economic consequences of the current pandemic, in the recent superstorms of the Caribbean, in the collapsing ice sheets at the Poles, including the third Himalayan Pole, and in the smog choking mega-cities in Asia and the Sub-Continent. As we started this book, fires were sweeping across our drought-stricken Australian continent. More than 12 million hectares, a billion native animals, vast swathes of National parks, iconic communities and local economies lost, perhaps forever – and we were only halfway through our expanding 9-month bushfire season. It is not 'the new normal' that gets so tritely tossed around. Normal no longer applies.

Also not an abstraction is growing and immediate mass economic dispossession in converging globalised crisis conditions. The scale of this pandemic depression is breath-taking in its interconnectedness, scope and depth. Its potential economic and political repercussions are illustrated in the post-GFC, anti-globalisation 'left behind' support for Trump in the rust belts of 'middle America', for Brexit in 'little England' and in the politically ambiguous Yellow Vests in France. This sense of dispossession will pale in comparison to the repercussions for billions in the less developed but interconnected sectors of the global economy. Global inequality was already trending to record levels pre-pandemic. Also on the rise were failing States, internal civil strife and dislocations, and cross-border migrations to escape geo-political, economic and climatic breakdowns. It is difficult to imagine that this will not accelerate as the current pandemic crisis and broader climate emergency unfolds.

YES, WE CAN

The inscription on Dante's gate is 'Abandon all hope, ye who enter here'. This book intends to contest that imminent doom.

There is a considerable degree of intellectual pessimism sweeping across our progressive left political culture. It would be delusional if it were otherwise. The task and timelines for real action look daunting and maintaining some optimism within our political spirit even more so. Much as we need to find a way to articulate a collective future worth having it would still only be a vision of hope and possibility. The immediate point (and urgency) is to *change the current political direction* and the future world it portends. We need to reconcile our still disparate voices to the political conditions as they are and where they are headed. Only then can we work out the best 'next steps' required for transforming a diverse coalition of opposition into a movement for democratically agreed change.

The forces of reaction are swirling. It is clear that deteriorating economic and ecological conditions will be fraught with perils associated with fragmenting, constrained and contested political choices. But within this vortex of looming despair lies the opportunity to create a new course.

The intensifying structural crisis of capitalist growth and climate emergency will demand a political solution to the growing political vacuum of a disintegrating liberal democratic capitalist centre. If the Left fails to fashion a transformational narrative that fundamentally challenges both business and politics as usual, then Ultra-Right narratives will continue to successfully cultivate support.

Imagining and shaping an aspirational collective future can only flow through galvanising the desperation of our unfolding economic marginalisation and ecological destruction. Ultimately it is only the collective political strength of our subordinate class majority to resist dispossession and entrenched power that will determine the political outcomes. And it is the precarious economic conditions and interests of working people that will bear the brunt of political failures. If there is a future worth having it must be born in this democratic crucible.

The fortressed moral bankruptcy, private self-interest and political hegemony of the possessors cannot be accommodated. That is the clarity of the navigational chart this Dantean passage requires. Much has been lost and there is no new ecological paradise that will be found. But no future at all apart from what we dispossessed make, *collectively*.

THIS BOOK

This book addresses the central question of mapping this collective destination home. Our core focus is to outline the political economy that underpins the science of climate change and ecological extinction. The book is primarily directed to a broad progressive audience, including especially climate action and anti-capitalist globalisation activists, who are already engaged in doing something about the ecological emergency and those concerned about what the future may hold should current projections be realised. Those readers who are still unsure about the science or want more detail may wish to read some of the works listed under 'further reading'.

The format of this book is fundamentally a reflective and perspective-based activist contribution to current global debate. We draw, often experientially, upon many diverse works and traditions across politics, academia, media and popular political culture. In other words, it has emerged rather organically from the course of our political and intellectual lives. As such, we have consciously avoided chopping the overall narrative with direct references and supplementary annotations. Key references are acknowledged in relevant chapter notes at the end of the volume (in the form of 'further reading' lists). By and large, however, the influences and origins of perspectives outlined are difficult to precisely identify and even more difficult to locate and accurately pinpoint. Apologies to the reader keen to head directly to source material. Also, special apologies to contributors who have shaped our expressed views and may feel short-changed in the acknowledgments. It is not deliberate or intended.

This book is overwhelmingly influenced by some 50 years spent studying, thinking, writing and campaigning in social movements, largely from within a Marxist tradition. The ghost of Marx flows throughout this book, hopefully in the positive way it is intended. It has been the primary filter and prism through which our diverse perspectives and somewhat divergent paths have been shaped.

For Rob this primarily translated into a long academic career in the social sciences, in particular 'green criminology'. For John, a more politically activist course beckoned. We have collaborated throughout these years, regardless of where our personal politics have roamed. It is reflective of current political conditions that we have now come full circle to focus upon this more expansive joint political and intellectual venture.

In our early student days Marx briefly became a reinvigorated 'flavour of the month' in social science circles in the Academy. It was customary to dive into the 'sacred texts' as both an intellectual end in itself and as the source of validation for contemporary arguments, even ones tangential to anything Marx may have contemplated in his time. This still remains a tradition within the anti-Stalinist revolutionary socialist movement outside the Academy, each remnant version still building or clinging to its unique political character traits with added inflections from Lenin, Trotsky and so on.

Marx once felt moved to sarcastically comment that he was not a Marxist. We don't think his view would have changed much over the 140 years since. We have consciously sought to avoid getting the reader, and ourselves, sucked into the vortex of the staid interpretive whirlpool that is Marxist scholarship. Noted in the end references are some particular texts that we think provide insights and signposts that lend analytic clarity to our times.

Marx truly lives across the years through the vision, method and passion that drove his scholarship. We hope this book does some credit to the very best of the intellectual legacy he has left behind. As for validation, we leave the reader to draw their own conclusion on the merits of the commentary and perspectives we outline. We hope it contributes to advancing the political changes we require to bring about the social and ecological changes we need.

Finally, a note on the book's structure. We intend the book to be read sequentially. It begins with historical and conceptual building blocks that describe capitalist political economy and hegemony. It highlights the current ecological crises and structural contradictions in global capitalism. It then moves to discuss more directly the political challenges these trends present and the demands they make upon the Left and more broadly environmental activists to re-shape our intervention around climate politics.

In reflecting on the consequences of an economic system grounded in continual and maximum exploitation of a finite natural world we argue that unsustainable growth *has to* come to an end. This is different from why it *should* come to an end, as if it's a moral, intellectual, rational or political choice. The corridors of power in the global economy are already reflecting the integrated transitional challenges flowing from unsustainable growth. But these transitions are shaped by their interests, not those of our global majority. Change is coming, the options are stark, and if we as a movement don't get politically ahead of this extinction curve then the future will be very bleak indeed.

Thus, while this unsustainable capitalist growth *has to* come to an end it doesn't mean an almighty political shove is no longer required. To the contrary, it opens the door to consideration of 'what is to be done?' in relation to the interests of the global majority in such circumstances.

An important aspect of this analysis is why a 'decisive progressive movement' in the advanced countries has failed to emerge in response to this most
transparent economic and ecological crisis. In part this reflects intrinsic
reform approaches that offer negotiated 'green deals', from our side and
theirs, within a model of eco-capitalism. To be sure there are many dedicated
advocates with the sincere hope that through this reform approach the
extinction curve can be flattened. We outline why this greener capitalism is
fundamentally a dead end for our movement, many of our global majority,
and the ecological conditions of the planet. From this critique, re-calibrated
approaches to our political challenges are advocated and new directions
identified.

2

BEYOND THE HOLOCENE EDGE

The Covid-19 global economic lockdown has created shards of light in the green gloom enveloping the planet. The brown haze over Europe has dissipated. People in the Punjab can see the Himalayas for the first time in 30 years. The before and after lockdown pictures from Delhi through to Bangkok, Beijing and beyond show an atmospheric clarity of possibility. But sooner or later economic lockdown will end. Business as usual is scheduled to resume as soon as possible. When that happens the shimmering oasis of a greener future will be revealed as a cruel mirage. The ecological extinction curve may show a temporary blip in the upward trajectory, maybe not. But its momentum, which has been established over several centuries, will not diminish without substantial political and economic transformation.

In this chapter we trace the scale and causal interconnections of the ecological catastrophe now unfolding. Humanity has reached a critical geophysical juncture in global climatic conditions. The 15 degree 12,000-year Holocene 'goldilocks' period is over. The Holocene marks the relatively stable interglacial warm band of 'recent' earth climatic history. This refers to the roughly 10,000 to 12,000-year *2-degree maximum temperature band* within which diverse human 'civilisations' firmly established and expanded.

This time frame, and in particular the temperature band, is cautiously generous. The historical data on temperature is established by (pre-thermometer) proxy methods. Most research on the Holocene marks the period from 11,300 years ago and the temperature band range as predominately more like 1.5 degrees, not two. A broad range and definition are warranted to avoid pointless boundary disputes distracting from transition challenges. Notwithstanding sharp episodic temperature spikes above and below the 15-degree average, the oscillation within the band has been about 0.75 degrees either side. Previous social evolution of the human species occurred under much colder average global temperatures.

Existing temperature increases place Earth at the Holocene edge and are projected to increase at an unprecedented rate and scale since complex societies developed. We have actually entered conditions comparable to the most recent interglacial warm period (Eemian, ca.125,000 – 115,000 years BP – before present time), which was some 1–1.5C warmer than the Holocene average, pre-industrial. As a consequence, ecological conditions that have been taken for granted across millennia have entered a perilous new territory. The climate science clearly marks us as being at the edge of stable Holocene climatic conditions. More ambiguous is collective clarity on how we got to this point and the full socio-economic and geo-political consequences.

WHY THE HOLOCENE CONTEXT IS IMPORTANT

The Holocene context sharpens the dimensions of the climate crisis we are confronting and the scale of economic, ecological and political/ideological challenges. For a start, episodic temperature spikes aside, the narrow Holocene band highlights the fragility of the global climatic conditions for production, reproduction and expansion of human society.

The narrow average temperatures of the Holocene band nonetheless mask considerable regional climatic variations, fluctuations and consequences. Historically, even small global average variations within the band have had dramatic impacts on human populations, social evolution and history – expansionary and contractionary – with localised societal overreach in production, consumption, population, ecological limits and terminal decline in a myriad of historical civilisations (e.g. the Mayans, the Anasazi). While we have historically experienced (essentially localised) extinctions and mass migrations before, they have certainly not been on the predicted scale of the present era.

The most dramatic context of the current crisis is that the 4 degrees shift from the last Ice Age into the Holocene was at the comparative geological sprint of 6,000 years. Pre-Holocene social evolution (mostly hunter/gatherer, often nomadic) occurred under conditions 2–5 degrees colder. The pre-industrial global average temperature of the eighteenth century of approximately 15 degrees was relatively average for the Holocene period as a whole. The current temperature is now above this average and has been accelerating since 1900 and particularly since the 1970s. At current pace we will accelerate through Eemian conditions within the next two decades. This will place us beyond the temperature band of glacial-interglacial cycling over the past 1.2 million years.

This foreshadows an equivalent 4 degree shift out of the Holocene will be at an interstellar speed of 100 years from now, by early next century. Moreover, we are hurtling past this Holocene edge with a population of 7.5 billion, not the pre-industrial 1 billion, and are heading for 10 billion by 2050. By this time, on current emissions projections, the average global temperature could be as much as 3 degrees above the pre-industrial Holocene temperature average and thus fully 2 degrees outside the Holocene band.

The most comprehensive and up-to-date compendium of the climate science, and most alarming given its historically conservative reporting structure, can be found in the latest Intergovernmental Panel on Climate Change (IPCC) reports. To these we might add recent and ongoing updates by the World Meteorological Organization, the United Nations Environment Programme, UNICEF, and the Lancet Commission, all of which are sounding the alarm on the rapid heating of the planet due to carbon pollution as well as providing detailed descriptions of the consequent climate changes and environmental disasters.

The full scope and nature of the Holocene Epoch heightens the significance of precisely what has happened since the advent of the Industrial Revolution – the causal driver of the so-called Anthropocene Epoch that is about to be officially periodised from 1950, reflecting the measurable science of the "Great Acceleration" in human impact globally.

The agreed (non-binding) target from the recent Paris Climate Accord (2015) is to limit temperature rise to well below 2C and to make efforts to limit the rise to 1.5C above the 15 degree pre-industrial average. This requires staying below 450 ppm (275 ppm pre-industrial, currently about 418 ppm). This is at best a 'practical' decision since there is no political stomach to even entertain the pretext of lower targets. In any case, this doesn't change the actual reality of heating and its consequences.

With only a 15 degrees average Holocene global temperature an increase of two degrees is significantly outside the stable Holocene band. When the global average was 13 degrees vast areas of north-western Europe were still under slowly retreating ice sheets. So, 2 degrees relative to 15 degrees is substantial, and the picture gets worse.

The Holocene band is projected to be displaced by a qualitative shift proportionately equal to that of the shift from the pre-Holocene largely nomadic hunter/gatherer period (approximately 2–4 degrees from the Holocene average). At current rates, this shift will likely occur within the next 30–50 years.

We are already at 1.1 degrees Celsius above pre-industrial levels. Notably, since the 1980s, each successive decade has been hotter than any preceding decade since 1850, and the trend is only accelerating. Ocean heat content reached record levels again in 2019 – which is significant insofar as the ocean absorbs over 90% of the heat trapped in the Earth system by rising concentrations of greenhouse gases. Two degrees of planetary heating is probably already in the pipeline at the current level of 418 ppm CO_2, given recent accelerations of land and sea ice loss and rising deep ocean temperatures. 450 ppm would make it a certainty.

Most popular/political discussion is focussed upon CO_2, which has been perfectly okay as far as it goes. CO_2 is the longest living of the greenhouse gases, which means that its use as a measure foreshadows the greatest global heating commitment in the pipeline across several hundred years. The safe level is estimated to be below 350 ppm, which still sounds like an accessible target for mitigation and possible reversal, thus preserving collective hope. Yet, with every delay in meeting required cuts to contain projected emissions levels below even 2 degrees Celsius, the greater the size of the emissions cuts required in the future. In other words, each delay now will demand even larger cuts later. At the moment we are still going the wrong way, led most emphatically and deliberately by the United States.

Moreover, as tipping points are passed and earth sub-systems increasingly go into positive feedback, more potent greenhouse gases like methane (20 times more lethal than CO_2), will also increasingly become a greater part of the total component. The CO_2 carbon equivalent (ce) of total greenhouse gases is currently over 500ce ppm.

Methane is short-lived compared to CO_2 and is currently increasing from anthropogenic sources (e.g. gas fracking, ruminate animal farming). However, the potential to massively increase the methane ratio as a greenhouse gas comes predominately from warming geological sources (frozen methane ice crystals, i.e. clathrates, trapped in deep ocean seabeds and in organic matter in the frozen tundra). These are not capable of being mitigated because of their scale. Crucially, as this happens, it will mark the transition to a self-expanding greenhouse dynamic, regardless of future emissions control efforts.

We are thus already in very dangerous territory and have no realistic means through which to significantly extract and reduce current unsafe greenhouse gas levels. The critical tipping points for the planet are inexorably being reached. Central planetary cooling mechanisms are either already entering positive feedback (e.g. summer melting of the northern ice sheet, reducing the reflective capacity) or set in course as irreversibly committed geo-physical outcomes (deep structure ocean warming, acidification and carbon sink overload).

The North Polar Region is heating at an accelerated rate (about double) compared to the rest of the globe. Siberia today is experiencing record warm temperatures in spring, well before the advent of summer. The thawing of methane-trapping permafrost appears imminent and unstoppable at the temperatures projected. The Amazon has had three crippling droughts this century and in the next couple of decades is poised to become a net contributor to carbon emissions rather than the massive carbon sink it has traditionally been. The accelerated razing of the forest for agricultural capital under the Bolsonaro government is exacerbating this exponentially.

Accompanying climate change are ever-widening destructions that threaten global ecological wellbeing. The loss of biodiversity, for instance, continues at a rapid pace and the principal pressures driving biodiversity loss are either constant or increasing in intensity. Among other things these pressures include the unsustainable harvesting of natural resources, the loss, degradation and fragmentation of ecosystems through land conversion for agriculture and forest clearing, and global heating, all of which are interrelated. Most deforestation is occurring in tropical forests with substantial biodiversity impacts, and net losses are especially significant in South America and Africa. Recent events in the Amazon and Indonesia, featuring widespread deliberately lit fires, makes the situation even more worrying.

In 2019, it was reported that around 1 million animal and plant species are now threatened with extinction, many within decades, more than ever before in human history. The ramifications are catastrophic. Meanwhile, the Covid-19 pandemic once again highlights the interrelated nature of these compounding problems. For example, air pollution both contributes to climate change and exacerbates health problems – fossil fuel-related emissions account for about two-thirds of the excess mortality rate attributable to air pollution. It is significant as well that over 400 excess deaths were attributed to smoke from the 2019–2020 bushfires in Australia, with thousands more admitted to hospital for heart and lung complaints and for asthma. Those suffering from air pollution are also among the most vulnerable to the coronavirus. There is thus a link between air quality and the damage wrought by Covid-19.

This example of the human 'extinction curve' is in fact made up of several moving parts: the increasing advent of pandemic viruses has been an *expected and anticipated* outcome of climate change stemming from global heating. In the case of Covid-19, its advent seems to have arisen from the increasing depth in boundary interface between globalised human economy and natural ecology, enhancing zoonotic disease transfers. The rate of zoonotic diseases (e.g. SARS, Ebola) appears to be increasing.

With respect to greenhouse gases alone, we are essentially out of time to stop catastrophic ecological change, irrespective of any extreme and urgent measures that may be taken from this point on to reduce further emissions. Of course, urgent action is barely on the agenda yet. The largest nation state polluters, aggregate and per capita, are still refusing to accept binding targets and are at best only still willing to talk about talking. Copenhagen, 2009; Warsaw, 2013; Paris, 2015; Madrid 2019 – one wonders about the content of the next theatrical performance of the climate action pantomime in Glasgow 2021. The Warsaw UNFCCC meetings, for example, coincided with the conference of the World Coal Association, overtly supported by Poland's Ministry of Economy. We should not be surprised, then, that global fossil fuel consumption *subsidies* increased by 50% over the three-year period to 2019, reaching a peak of almost US$430 billion in 2018. Hollow rhetoric emanates from nation-states that still actively support and represent the vested interests perpetuating the very thing which they profess they want to stop.

Should there eventually be 'urgent action' to curtail the pace of global heating, in particular that generated by fossil fuelled energy production and consumption, then we will immediately have to confront the end of global dimming. Global dimming (the reduction of direct solar heating as a consequence of atmospheric pollution, e.g. aerosol particles) has likely dampened the increase of overall planetary temperature, and certainly regionally, by perhaps 0.25 of a degree. In short, by adding the solar heating effects of substantial aerosol pollution reduction to the existing pipeline of committed heating from greenhouse gases, we are most likely already confronting an imminent global temperature increase of 20% or more above the Holocene average. Aside from geo-engineering techno-fantasies to mimic global dimming, it is near impossible to reconcile the projections with a fathomable solution to a very dire increase in global heating.

THE POST-HOLOCENE TRANSITION

Human civilisation in the Holocene has rested upon the dual capacity to predictably draw upon nature's free assets and to generate a continuing social surplus above the requirements of the reproduction of labour power required for its production. Without this stable *social surplus*, the development of society-building activities (such as specialist expertise-based art, music and crafts), large construction projects (the pyramids and other iconic structures), and of social layers 'liberated' from the task of productive labour for their own necessary reproductive consumption (ruling classes and religious orders), would not have been possible.

Thus, Holocene era climatic stability exhibited the dynamic foundation for the development of agriculture, technology, population growth, structures of governance, cultural activities, permanent dwellings and so on. It secured the six pillars underpinning the survival and continuity of collective societies: *water, air, food, energy, shelter* and *security*.

This period also provided the underpinnings of class society and the institutions of social control and armed security protective of the appropriation of the social surplus. Various modes of production emerged – the ancient based upon slave-ownership; the feudal grounded in aristocratic ownership of land; the capitalist centred on exploitation of labour – each with its own form of appropriating surplus, ruling and subordinated classes, and systems of coercion and consent. Transitions from one mode of production to another, for example, feudalism to capitalism in Europe, took centuries.

Human history is punctuated by periods where civilisations have radically transformed and/or crumbled through the advent of unpredictable natural events, self-inflicted environmental folly, or engaged in a war of survival to secure, enhance or defend their material conditions of existence. The overall pattern, however, is one of relative continuity in the development of human civilisations across the globe.

Whereas *climate science* marks us as being at the edge of transition beyond these historically stable climatic conditions (there is near universal consensus on this), more ambiguous is collective clarity on how we got to this edge (with diverse explanation ranging from unbridled population growth through to specifically human over-consumption). This seems as surprising to us as it is disturbing given the stakes involved. From our perspective this needs to be scrutinised forensically in ways that point to the more distinct and specific political economic markers that are driving change in the Earth system.

Accordingly, we are not convinced about the merits of designating 'our' contemporary convergence period of earth science and human history, as the new Anthropocene Epoch. There is now a formal proposal in the geological community for the recognition of the Anthropocene in the Geological Time Scale, with a recommended start date in the mid-20th century. However, it is the 270-year-old Industrial Revolution in which transformational human impacts on global earth systems became qualitatively more pronounced. We understand the political aspiration to contest denial of human-induced global heating by reference to the Anthropocene. Nevertheless, whether as critique about 'our' destructive capacity and effect on 'natural' earth systems or as a philosophical/ideological objective of exploitation and technological control by neo-liberal, free market fundamentalists, there is a distinct ideological component to the Anthropocene Epoch designation. These formulations tend to mask the important fundamental relations that go to the heart of the matter.

Tying 'the Anthropocene Epoch' to the start of 'the Industrial Revolution' would be ironically appropriate as both phrases rest upon a shared fiction and shared silence: that it is human technological development that is the curse and/or the solution rather than the economic and social relations within which it has been historically embedded. It was never just an 'industrial revolution', as if it sprang up ontologically separate from the capitalist system already in existence. Technological development throughout human history has always been bound up with the existing mode of production of social and economic life and driven by its dynamics. The industrial revolution was no different, nor is the so-called post-industrial revolution of our contemporary historical period. These technological dimensions have not developed as if cocooned from capitalist profit growth imperatives, class interests and relations of power. They stem from and are embedded in these.

There would be little dispute if we collectively acknowledged that it is the origin and development of the *capitalist* industrial revolution that underpins sharply accelerated climate impacts and degradation of the ecological conditions of human development generally. It is important to realise, as well, that although climate impacts were only really measurable for CO_2 and temperature years later (1900s onwards), ecological degradation was already well underway in the prior centuries of development of European merchant and commercial capital. Colonisation, natural wealth transfer and the post-feudal capitalisation of agricultural industrialisation (e.g. the sheep/wool industry) transformed vast sections of the natural world in Europe and spread globally (e.g. sugar plantations). When the capitalist industrial revolution got under-way, however, this degradation increased exponentially. The post-Holocene transition rightly should be marked as structurally commencing from this point.

Humanity as a whole, now and historically, stands in relation to the natural world as bounded and mediated through the social and economic relations we have constructed with each other. It is through the prism of various modes of production and reproduction of economic and social relations that humanity's historical and contemporary relation to earth systems and so-called 'crises' must be framed and evaluated. There is a need to resist any conceptualisation (such as Anthropocene Epoch designations from both 'left' and 'right') that masks this historical evolution throughout the Holocene, and its transition dynamics.

The development of Holocene epoch civilisation (i.e. post-nomadic) already marks the qualitative beginning of anthropogenic transformation of the human-lived environment. It is the *scale* that is in question, whether it is local, regional or of earth system impact. To the extent that there is a human effect impacting upon, controlling or transforming earth systems it is ideological hubris that 'we' could, should or already have (even in a negative destructive way) only now taken over effective management/transformation of earth systems.

At one level, 'we' are still just doing what 'we' have always done, even if the scale is now vast relative to previous localised examples (e.g. Easter Island).

On the other hand, our collective human grasp of the complexity, systemic interconnectedness and causal feedbacks of the natural world is feeble at best, at least as presently institutionalised in government and private-sector ini-tiatives. Just because we are toileting in our own drinking water and are happy to filter and do it again, or can build canals, reservoirs, dykes or desalinisation plants to subdue seasonal variations in rainfall and storm intensities, doesn't mean we have control of hydrological and oceanic cycles (e.g. the demise of Angkor Wat or future perils of the Netherlands and Bangladesh).

The 'technological exceptionalism' of the *human species* embedded in the notion of the Anthropocene is eerily reminiscent of Fukuyama's ode to Western liberal democracy and the 'end of history'. It suggests that we have entered a brave new world where the dynamic motors of ecological, geological and evolutionary 'pre-history' have been suspended and/or superseded and that 'we' are now in charge of Mother Nature and 'our' future. Avoiding this (latent) ideological hubris is important, even as critique. If not, we suppress the determinants driving global heating (specific social relations of production and accumulation) with their expression (particular means of production, including technology). Emphasising the latter just privileges and leaves uncontested the continued dominance of the former.

Fear of survival is the only missing bit that separates the ideological space between 'right' anthropogenic exploitation/control of the natural world and a 'left' anthropogenic aspiration for more harmonious environmental balance based on renewable energy and re-regulating capitalist excess (perhaps with some legally enshrined ecological and species rights thrown in for good measure). The shift to now embrace the prospect of geo-engineering, out of necessity, in mainstream government and business circles is a case in point, as is talk by some progressives about the saving graces of nuclear energy and genetically modified (GM) food. The recent first world 'public shift' from vested interests denying and minimizing future global warming catastrophe to '(oops) it's too late to stop it now' underscores and amplifies the ideological and political trap. Technology is seen as both problem and saviour – by both sides.

The increasingly farcical debate with neo-liberal ideological deniers of human-driven climate change like the Heartland Institute and the Koch family will soon be over. The US military/industrial complex has been war-gaming catastrophic global warming since early this century (i.e. under the neo-liberal regime of Bush-Cheney). Political wrist-slashing over the shattered environmental hopes for Obama's social democracy is to completely miss the point and, more importantly, the historical moment.

The self-trapping ideological/political risk is embedded in conceding the 'dark victory' of fabricated neo-liberal climate denial. The real character of this dark victory is 'us' continuing 'our' historical ideological complicity in mythical super-structural reforms of the mode of production that got 'us' here (as if it is simply a matter of the right regulatory policy and/or the right technology). Those of us in the bloated 'middle classes' of the advanced capitalist countries largely still remain shared beneficiaries of its relative material comforts. Our fortressed and quarantined air-conditioned spaces and affluent buffers of disaster re-construction still leaves a dark defeat for the billions who aren't beneficiaries. Moreover, it foreshadows a similar exclusion for the still privileged millions of 'us' who eventually won't feature for buffering as our looming future unfolds.

This highlights an additional and perhaps more politically important component of ideological obfuscations such as the Anthropocene designation – the ubiquitous 'WE are all in this together' mantra (as in shared complicity, perils and solutions in discussions of the post-Holocene transition). Perhaps 'we' trips more easily off the tongue when one has an affluent stake in the system that grants one's relative prosperity. Perhaps it flows from political identification with a social democratic capitalist reform perspective to win social/economic concessions and occasionally even progressive government in expansionary capitalism's crisis periods. (Like Obama, no 'they' – the Roosevelts, Kennedys, Clintons, Whitlams, Blairs – couldn't deliver for us!) Perhaps it simply reflects a professionalised ideological location within the various apparatus of capitalist state power. 'We' is a hollow curiosity to the global majority outside the access and privileges of these vantage points.

The 'we' also implies that a middle ground exists within a class divided society, that consensus is possible across competing vested interests, and that the preferred political pathway is to win everyone to 'our' side. If we are all 'we', then our interests must be shared and universal rather than structurally contradictory and antagonistic, requiring fundamental social transformation for resolution.

Even if we grant that 'universal human interests' exist in the sense that certain threats affect us all regardless of class, gender, ethnicity and/or geography (for example, the threats of nuclear holocaust and ecological degradation), this is fundamentally different from understanding that it is class dynamics that are causing these threats. Specific class interests, manifest concretely in the actions of capitalist governments and transnational corporations, means structural unwillingness to implement or enact strategies and policies that would, in effect, protect these apparent universal human interests. Acknowledging this, places competing class struggle and conflicting interests at the core of climate politics.

Human civilisation inherited Holocene natural world conditions and flourished, and 'we' have trashed it beyond recovery within any geological time frame that will matter to us and our descendants. The future is only 'uncertain' for us as a species, but certainly grim if not already dire for vast numbers of us. 'WE' are not in control as there is no 'we' in reality as this scenario unfolds.

Divergent ownership, control and access to the levers of economic, technological and political control have shaped the contours of Holocene civilisation to this transition point. That is the evolutionary universal of our Holocene human civilisation and how we got to this perilous convergence in the first place.

Whether as productive subjects in tribalised, feudalised, nationalised or globalised structures and institutions of class power, 'we' is actually 'them versus us' in our very civilised human code. Unless there are dramatic socio-economic transformations, any newly designated post-Holocene 'Anthropocene Epoch' could be a very short epoch indeed for most of us. But for the globalised capitalist ruling class (and the down-sized minions, mercenaries and productive subjects they may still require in future) their affluent bubble is unlikely to burst anytime soon, Covid-19 notwithstanding. At the very least, this bubble burst will not happen without a stark transformation in business and politics as usual.

A GLOBALISED CAPITALISM

We argue it is overdue to clearly acknowledge that it is the origin, development and character of the capitalist industrial revolution that underpins sharply accelerated climate impacts and the degradation of the ecological conditions of human social life generally. This is largely masked in charting the history and core fundamentals of this crisis, even by many of the most dedicated advocates for climate action.

Our emphasis here is on 'capitalism', at the core of which is private ownership and control over the means of production and appropriation of the collective social surplus – and indeed the very material six-pillar basis of collective social life itself such as clean water and air.

What we mean by 'capitalism' is not just rhetorical condemnation of system evils. We all know that capitalism is 'bad' for the environment (as is

evident in the signs, speeches and chants of climate striking protests). However, we risk tendencies to blinkered and unexamined pejorative trivialisation
within our movement about capitalism. Rather, for us, capitalism embodies
social relations that must be unpacked and analysed precisely because the
enhancement of political perspective and transformative strategy demands it.

Like others, we note that the post-Holocene transition structurally com-
mences from the capitalised industrial revolution from 1750. Existing
ecological degradation increased exponentially from that time. Since these
beginnings, capitalism is now THE global mode of production, both in length
of geo-political reach and in the depth of integration of antecedent modes of
production and the masses of people immersed within its structures.

The inter-imperialist expansionary wars of territorial acquisition/colonisation
(to economically secure national capitalist production, consumption, distribution and exchange) of the sixteenth to nineteenth centuries have been progressively displaced over time. Now multi-regional capitalist blocs as well as globally
internationalised governance structures accommodate the unfettered requirements of global capital investment flows, endlessly expanding commodity production, and the realisation of capital investment through equally expanding
global consumption. Capitalism as a system has subordinated all other forms of
production and consumption, displacing and marginalising alternative ways to
satisfy social need.

Its most evolved organisation expression – the transnational corporation –
dominates the global landscape. Transnational companies are now dwarfing
some countries in regards their income, resources and people. Criticised by
early capitalist political economists like Adam Smith for its potential to
pervert market competitiveness, the warnings of concentration of capital and
monopoly have proved prediction. Power and wealth are now concentrated in
conglomerates of huge horizontal and vertical density across an integrated
array of commodified sectors (e.g. food, water, energy, transport) and
controlled by individuals with scarcely believable personal riches.

The billionaire is now accorded the status of folk hero in commercial
media, itself owned and controlled by monopolies. Private profits are treated
as public goods and in the public interest – such is the mantra of the late
capitalist think tank, the CEO and spin machine merchants, as is evident
in the proposed billion-dollar bailouts of big business in the United States due
to the coronavirus pandemic. And, even if 'we' don't all believe them, we
believe that it is the natural order of things or at least is unassailable. How

else can we explain that still, today, over 30 private-sector banks such as Wells Fargo, RBC and Barclays from the United States, Canada, China, Japan and Europe have together channelled US$2.7 trillion into fossil fuels between 2016 and 2019, that is, *after* the Paris Agreement was adopted. In conjunction with outright government subsidies, the support for dirty industries has continued unabated.

Globalisation, capital concentration in ever fewer hands and transnational corporations highlight the most obvious distinguishing feature between capitalism now compared to the capitalism of the classic imperialist period. Then, the capitalist mode of production (CMP) was increasingly dominant but there were still vast sections of the globe in which pre-capitalist relations of production and consumption prevailed. Over a 500-year period, these forms of economic organisation were geo-politically integrated into the orbit of capitalist domination through territorial conquest, often in the course of competitive struggles between imperialist nation states. Military dominance both determined and flowed from economic and technological dominance. The most successful capitalist states supplanted older and/or less competitive imperialist/colonial orders. This 'classic' form of Imperialism, that between capitalist states for territorial expansion/direct political control of the non-capitalist and underdeveloped sectors of the world, was largely ended by the conclusion of World War II and de-colonisation.

There is now no part of economic activity across the globe that is not directly shaped by the capitalist mode of production (CMP). Even where that CMP domination is still in its early stages of integration and transformation – as in the case of remnant peasant and petty-commodity production – there is no aspect of pre-capitalist economic activity that is not now substantially dependent on the consumption of at least capitalistically produced goods and services or exchange into capitalist-determined markets. This marks a distinct 'maturation' phase of the CMP. It signifies a territorial dominance of the CMP across every corner of the globe. It doesn't suggest a limit to the timescale of continuous expansion or even whether it has peaked in its historical evolution. That has yet to be determined. It also signifies one other crucial aspect.

The inexorable expansion of the CMP is now competitively bound within its own parameters and its own internal contradictions and relations: horizontally between national and international capitalist competitors and vertically between various types of capital and the mass of global labour at its disposal.

Capital now owns the planet (so to speak) but it also now owns the consequences of capitalist expansionary imperatives and the crises they engender. But it is no longer partially globalised post-war American capital that solely or even substantially owns the planet. After 30 years of genuine globalisation capital is now both multi-nationalised and internationalised. This is part of the economic and geo-political dynamic that needs to be examined.

Importantly, capital globally also owns the ecological contradictions and consequences engendered by endless expansion. How and to what extent this will shape, constrain and undermine the very conditions for continuing expansion of the CMP in this conjuncture is a central set of questions. It bears directly on the question of capitalist hegemony, its structural contradictions and crisis dynamics.

We conclude that climate catastrophe is thus unfolding in the midst of a critical transformational juncture of global capitalist development. The 30-year boom of capitalist globalisation integrated and expanded manufacturing and financing across the hitherto Stalinised and underdeveloped nation states. As reflected in the global financial crisis (GFC) of 2008, and now today in the Covid-19 economic crisis, the imperative for continuous capitalist growth has reached key structural impasses. It is a crisis of and for the continued expansion of the CMP in this post-Holocene transition. By extension, the climate crisis and ecological collapse are, simultaneously, an integral component and antagonistic consequence embedded in the very fabric of this historical conjuncture in globalisation.

Subsequent chapters in this book spell out the cascading economic, ecological and geo-political ramifications.

3

THE EXTINCTION CODE WITHIN THE CAPITALIST GROWTH PROTOCOL

The historical capitalist origins of the post-Holocene transition are central to clarifying the catastrophic character and dimensions of global heating and ecological collapse. This chapter focuses on the inner dynamics of the capitalist growth protocol that is driving this integrated post-Holocene catastrophe and its seemingly politically stubborn inevitability.

CAPITALISM REQUIRES GROWTH

In its fundamentals, extinction-level ecological degradation and transformation are specific features of capitalist economic growth imperatives. These are inseparable from the cyclical expansionary and contractionary dynamics in accumulation, and divergent class structures of power and interests.

There is no shortage of commentaries that argue this integrated ecological crisis is a consequence of industrialisation, unregulated capitalism, neoliberalism, free-market fundamentalist ideology, extensive fossil fuel use, excessive consumption and/or population explosion. What is striking about these is that the problem is not necessarily equated to capitalism and its core growth imperative per se. It is features of the system, not the system itself, that tends to be targeted for criticism.

The growing conundrum is that even the everyday Jo on the street intuitively grasps (like the School Strikers for Climate) that an endless growth economy is at odds with finite resources, that an endless expansion of consumption is required to absorb an endless growth in production, and that endless expansion of production and consumption means, if nothing else, endless by-product expansion of waste from planned obsolescence and pollution, even with diligent re-cycling. Also, intuitively grasped, is that the rich invest in order to get richer.

The key puzzle demanding explanation is why anyone thinks that an economic system based upon endless growth on a finite planet, and a tiny rich minority getting richer relative to the vast social majority, somehow will not be punctuated by recurring crises and systemic collapses – economic and ecological. This is the great myth/silence that emanates from the corridors of power and their spin merchants. We are sleepwalking to catastrophe pursuing 'endless growth and affluence for all'. It's a fairy tale that school children openly and publicly acknowledge in their social media posts, rally speeches and placards. Yet it's a fairy tale that seemingly infects many seasoned reform eco-capitalists: that a *green* capitalism, or a *no-growth* capitalism, or a *'we are all in this together'* more altruistic and benign capitalist system and ruling class, is possible.

There is an extinction code of exploitation that lies at the core of contemporary capitalist economy. This capitalist relation of exploitation with regard to labour and the natural world is poorly understood or acknowledged. Partly, this misunderstanding flows from pejorative connotations of the term 'exploitation'. Yet, this form of appropriation of the social surplus remains central to understanding how capitalism as capitalism actually works.

Also a factor is (possibly convenient) superficial universals relating to class structure, inequality and ecological degradation across previous historical periods and forms of economic organisation (from the ancient Holocene world through to post-capitalist Stalinisation). After all, why focus on capitalism as the primary problem when exploitation and ecological degradation have been features of our entire human history and across different forms of socio-economic organisation? The refrains are familiar: 'Just look at the ecological catastrophes of the old Soviet Union and the People's Republic of China'! 'Humans have always treated the planet instrumentally'! Such cries are not uncommon in both extended commentaries and short catchy tweets. In either case, our view is that the claims are simplistic and politically miss the main point, deliberately or otherwise.

The exploitation of labour and the natural world has not been confined to the CMP. However, the CMP relation of exploitation is historically unique and underpins why systemic economic crises and ecological calamities are endemic. It also underpins the institutional fabric of class interests and power that are indispensable to its preservation (as we discuss in the next chapter). By extension, these capital-centred forms of institutional hegemonic power are thus the central obstacle to any transition to new relations of economy and ecology.

CAPITAL AND CAPITALISM

Capitalism encapsulates and is defined by growth for its own sake. This imperative hasn't always existed throughout the economic history of societal development in the Holocene. Moreover, it hasn't always been the case that increasing the production of stuff or the accumulation of money have been ends in themselves. Nor has it always been the case that plundering the natural world regardless of the consequences is the core human social ethos (contrast, for example, the harmonious relationships between Indigenous peoples and their environments). These are evolved and inherited historical choices. They reflect historical relations that have been made, altered over time, and can be re-made.

Capital itself predates the 270-year-old CMP by thousands of years. It is created by labour. More precisely, capital is formed from surplus labour. This is labour that creates a social surplus beyond that which is socially necessary to maintain the conditions of existence of the labour that created that surplus (that is, the amount of labour required to feed, clothe, shelter and ensure the health of an individual over time). How social surplus, and thus the surplus labour that created it, transforms into capital has a long history and has taken a number of historical forms.

Earlier pre-capitalist modes of production largely consisted of agricultural and petty-commodity craft production (leaving aside Indigenous communal forms of production). Surplus was often initially a by-product of land-based peasant and/or slave labour. This social surplus labour (appropriated from peasants and slaves by owners) translated into goods or (eventually) money to be exchanged by the direct producers or owners of surplus product for necessary/desired goods produced by others. Historically, this started as predominately just a mutually beneficial, simple commodity exchange between direct producers (e.g. surplus wheat for surplus pork or tools) and just for their direct consumption. This exchange of different products (their quantity, quality and availability) was roughly equalised based upon the amount of labour respectively required to produce them.

Over time greater specialisation in the division of labour transformed surplus product, as a by-product of direct producers, into the purposeful focus to produce a particular commodity for market exchange to obtain a range of goods and services produced by others. These commodities contained both a use value (otherwise no one would buy them) and an exchange value representing the average socially necessary labour that was required to produce them. If labour creates more value in the form of commodities than it is required to maintain or pay that labour, then it creates a surplus value.

Before the CMP the principal basis for expansion of capital was based upon this market exchange of the surplus value of petty-commodity production. Commodities were exchanged via money as an efficient medium of market exchange and as a store of surplus value. The more efficiently and cheaply you could bring commodities for exchange on the market relative to others (through production or trade) then the more your capital wealth grew in tradeable commodities or the money equivalent. At its immediate pre-CMP historical height, this created a great deal of concentrated merchant and commercial capital generated through increasing global trade in commodities generated by several hundred years of European imperialist colonial expansion.

This growing concentration of merchant and commercial money capital became the structural antecedent for the historical transformation in Europe from feudalism into the capitalist mode of production. Money capital merged with developments in science and technology to create the industrial revolution, namely, the mass domestic production of commodities for market – locally/nationally and for export.

The capitalisation of agriculture and consequent transformation of feudal land to single commodity production for the market (e.g. sheep, cattle, grain) drove the bonded peasants and serfs from the land to create a mass disenfranchised labour force needing to find work in the emerging industrial factories in the cities, and in the coal mines powering these industries.

The CMP is thus distinguished by three core features from pre-capitalist modes of production:

1. Money capital is directly integrated with technology and the production of commodities to create *industrial production capital* (i.e. capital investment in machinery that allows mass production of goods);

2. Production labour is divorced from any direct connection to the means of production (for example, the peasant on the land had this connection, the proletariat does not). A class of *wage-labourers* is created whose only recourse is to sell their labour-time to capital in order to reproduce their conditions of existence. Labour itself becomes a commodity under capitalism. It is only from money wages that labour can buy the material goods required to live; and

3. The *production of commodities is generalised*. The sole purpose of capitalised commodity production is to create more money capital through *direct* expropriation/exploitation of the surplus value produced by labour in the commodity production process (the owners pay workers less than the total value of the things they actually produce). The more the commodification process can be extended, the more money capital and profit can be generated.

Within the CMP, capital thus socially appears to us in two forms: as money (which stores the exchangeable surplus value of commodities created by labour) and as a physical commodity, asset or means of production purchased for its use value to produce goods for exchange (e.g., factory machinery to make cars or furniture).

Not all money or physical assets are capital. As we know from everyday life, whatever money we may have isn't capital if we use it to buy food to eat or go on a holiday and not every physical asset we own, like our car or furniture, is an investment to make us richer for having bought it.

Therefore, *capital* only really exists when it is money or physical assets that are active or 'in motion' to make more capital – that can be utilised to make more money than we spent to start the process.

Capital then is both a process and a social relation. It has to be set in motion to create more capital; otherwise, it just sits there unproductively or worse begins to degrade (e.g. from inflation). It can only be set in motion to expand by labour and, specifically, by labour that creates more value than is used up in the costs (like wages) required to initiate the production process.

Pre-CMP, merchant and commercial capital grew from taking a profitable cost deduction from its circulation of the surplus value of petty-commodity production (domestically and imported). In other words, profits were made by bringing commodities to market for realisation in exchange for more money than it took to purchase or extort/swindle these commodities from direct producers and the costs required to bring them to market. The British and Dutch East India Companies excelled at this and became the first forms of Multi-National Corporation at the start of the seventeenth century.

Within the CMP material goods and services are directly produced as commodities, from the very beginning, in order to make a monetary profit (rather than profit being primarily from commodity circulation). Goods and services are not invested in first and foremost to satisfy and enhance social needs, but for the promise of financial return. While socially produced and organised, this process rests upon private ownership, control and appropriation of the material wealth accumulation process. A *commodity* is fundamentally valued for its *profitable exchange* on the market. While water is vital to human life, for instance, and thus has intrinsic use-value, bottling this water and selling it on the market translates this use-value into exchange-value, and thereby a source of private return or profit.

Generalised commodity production reflects that making commodities for market exchange is the central form in which existing wealth can be preserved and expanded. This expansionary imperative for private capital accumulation is foundational to the CMP.

If money or means of production capital are left idle, then their value degrades. The private investment required to produce socially needed goods and services *only* occurs if private profit is considered possible and is ultimately realised in exchange by those with money capital to invest. Otherwise capital is not invested, as the purpose/motivation of further investment by owners of capital evaporates.

Even more problematic is if invested capital is destroyed as unsold goods or services. If the latter occurs, this is a crisis for those who own capital. It is a result of the over-production of commodities for effective market demand. This often reflects an existing crisis for those who normally require but cannot afford these commodified goods and services to maintain their conditions of existence, such as with the mass unemployment of the current pandemic crisis. It certainly is also an immediate crisis for those whose employment is dependent upon the ongoing viability of the enterprise, such as is confronting workers engaged in mining what will eventually be stranded coal assets.

EXPLOITATION, SELF-INTEREST, EXTINCTION

Labour is systemically bound to this capitalist growth imperative. If the labourer can't sell her/his labour time for money to buy the commodified stuff they need then their existing conditions of life instantly degrade or are threatened with extinction. In a wage-based economy, unemployment and underemployment have direct impacts on life circumstance. This is exacerbated in contexts where state welfare support is at or below very base subsistence level or is unavailable altogether. The Covid-19 pandemic has brought this home to millions worldwide, instantly.

Meanwhile, if the capitalist cannot make a greater profit from employing production labour compared to their competitors, then the enterprise eventually fails, and the worker's employment ends. It is a race to the bottom when it comes to wages, exploitation of labour, and company profit levels.

This competition compels the capitalist to either directly reduce wages to the lowest amount possible or to decrease the relative wages bill as a proportion of the total cost of production. This can be done by lengthening the workday as much as possible, or compelling people to work harder, or employing vulnerable people (like undocumented immigrants or children) at cheaper rates of pay. This was very prevalent in the earliest phases of the CMP and continues to this day in certain sectors of the advanced economies (particularly agriculture) and in the emerging areas of the global economy (particularly textiles).

Most prevalent however, and often in conjunction with the above measures, is buying machines to replace existing labour to maximise the productivity of the remaining employed labour – in other words, create as much surplus value as possible from as little labour as possible to transform into as much profit as possible.

In the CMP, then, the social relation between capital and labour is a relationship of exploitation. At its core this has nothing to do with pejorative notions of fairness, or lack thereof, or that a particular set of wages are below the social average of wages in a sector, nationally or internationally, or that some workers are made to work harder or longer than the socially acceptable average.

The mantras of 'a fair day's pay for a fair day's work', or 'a living wage', or a publicly funded 'social wage' supplementing direct money wages, are social constructs based upon the legal-juridical acceptance of the private ownership and control of the means of economic production. It rests upon the enshrined legal and moral right for owners of capital employing labour to accumulate a private profit from the enterprise.

Labour employed by capital is exploited even if that labour is highly paid (e.g. tech labour in Silicon Valley) compared to the lowly paid (textile labour in Bangladesh). It is also immediately dispensable, as shown by the widespread lay-off of airline staff (from top to bottom) during the present coronavirus pandemic.

Exploitation in the CMP thus reduces first and foremost to this: *the more value producing productivity that can be squeezed out of purchased labour time, the greater the rate of exploitation of that labour time.* This relation of exploitation of labour is thus the essential mechanism of the capitalist growth protocol. Increasing the rate of exploitation of labour in production is the foundation for wealth creation and expansion and the fuel that motors the CMP.

It's important then to distinguish the *fundamental* character of capitalist growth from its often ideologically entangled expression in job creation, technological innovation, and improvement in the material conditions of life, especially for the most economically marginalised. Capitalist growth is about the unrelenting expansion of private capital and personal wealth for its owners. The rest is, at best, a necessary method or incidental by-product to that end, insofar as commodities without any use-value market demand (existing or through creative marketing) would remain unsold.

The reality behind the spin is that capitalist growth is more about the creative competitive destruction of existing jobs, technologies and the established material foundations that sustain our social life. This is part of the built-in imperative.

Capitalism must continually expand production and consumption. This involves new markets for capital investment, and new products for consumption to realise profits. Capital doesn't get invested if the owners think they cannot profit. More importantly, invested capital has to be reproduced as returned capital, including profit. Commodities have to be sold. Capital has to have *realised* growth or die. This is the core mission that lies at the heart of the capitalist system. The economic fragility of the edifice was there for all to see during the GFC and more recently is being witnessed in the unfolding economic repercussions of the coronavirus pandemic.

Also transparent is the core nature and priorities of the capitalist 'democratic' state when the system is under threat. Billions and trillions magically appear to prop up failing enterprises and stock markets. For public health services and environmental crises (even protection of iconic species such as the Koala) not so much. In this regard, it is instructive to watch where the money flow from governments is going during the present global pandemic, as they struggle to prop up 'free market' enterprises while simultaneously confronting the health-generated collapse of labour markets and subsequent under-consumption of capitalistically produced goods and services.

It is in the private self-interest of the possessors of capital to grow their money capital and private wealth, regardless of the consequences. The rest is obfuscating ideological noise. It is pursued unrelentingly in relation to other capitalist competitors, in relation to its employed labour, irrespective of any social or national interest, and regardless of any existing or future ecological or generational consequences.

In the case of climate inaction, there is no shortage of reactionary ownership interests trying to cling to their existing and ageing technologies and industries (like coal and other fossil fuels) in the face of unequivocal science and looming catastrophe. It's only the vested self-interest that matters. But this applies to our green capitalists pursuing commodity production for transition into our emissions-free, post-fossil fuel future as well. It is as much a part of competitive elimination of labour and 'old' commodity industries through injecting new technology and maximising exploitation as any other sector of capitalist economy. Just ask Elon Musk's workers at Tesla.

As indicated, this capital-labour relation of exploitation extends to the relation of privately-owned capital to the natural world. The monetised, privatised, and competitive drive to maximise exploitation of nature has been completely and utterly devastating for the natural environment with the advent of the CMP. Just as there is the drive to maximise profit by extinguishing as much labour as possible from the commodity production process, so too there is the drive to consume (and thus extinguish) as much commodity production value from nature as well. Whatever exploitation of nature occurred pre-CMP, the exploitation in relation to generalised and mass commodity production was now on steroids.

The earliest, most obvious and comprehensive examples of this exploitation have been in the capitalisation of agriculture and extractive industries of forestry, fisheries and mining. Soil nutrient depletion associated with the transition to monoculture commodification was already a pressing problem in Europe in the late eighteenth century (indeed, the importation of guano from South America and the use of ground bone of the dead from the Napoleonic wars were widespread by the middle of the nineteenth century). Hiding behind the bucolic landscape paintings of John Constable was the increasingly abject suffering inflicted on sentient beings by the capitalising meat and dairy industries to bring mass commodified protein to market. The clear-felling of native forests for agricultural production and wood products reached industrial scale at the same time, basically eliminating the old growth forests of Europe. Industrial-scale fishing fleets drove some whale species to the brink of extinction (for oil and meat) by the late nineteenth century, as well as initiating the first of several steep declines in the abundant North Atlantic cod (a source of cheap protein to maintain the industrial working class). The famous London fog of the nineteenth century (covering some 20% of days per year) primarily came from Londoners burning Yorkshire coal.

As the compounding growth of the CMP has continued so too the extinction of the natural world has compounded. Now we are not only in the midst of the sixth great mass extinction of species in Earth history but on the absurd course for self-extinction as well. Maximum exploitation of the natural and social world has inextricably been bound together. The 'free gifts of Nature' and the environment generally – consumption for commodified production and as a waste dump for the end point of the commodity production and consumption process – has only been free for some. It continues to be a very socially and environmentally expensive and un-costed extinction process for most of us. But for private capital expansion, the profit flow has been immense.

While constituting the foundation for generating historically tremendous material wealth, the CMP is a narrowly defined and essentially monetised private wealth expansion imperative. This may be rational in its immediate self-interest for those possessing capital but is irrational and crisis prone systemically. It is certainly totally irrational from the vantage point of the collective interest of those being exploited and dispossessed, socially and ecologically.

There are limits to everything. In the crises that emerge when those limits are reached there are political opportunities everywhere to fundamentally change course. Capital, to coin a phrase, has not only created its own gravediggers in the dispossessed working class but has been creating its own barren wasteland for the grave itself.

The challenge is finding the democratic collective will and grasping the recurring crisis opportunities on hand to turn off the life support system of exploitation that keeps the private money flow going. Capitalism serves up plenty of these occasions.

STRUCTURAL DYNAMICS AND CONTRADICTIONS

Capitalist growth, for all its dynamic imperative to expand, does not display a uniform, linear trajectory. There is a core dynamic of regular cyclical crises of over-production and under-consumption of wealth-bearing commodities in capitalist economy. Expanding ecological collapse amplifies this cycle of 'normal' accumulation crises of capital and is, in turn, reciprocally amplified. Climate change and ecological collapse are and will be causally forcing dramatic structural transformations in global capitalism.

In particular, there are crises in maintaining and reproducing the standard of material conditions of life for wage labour. This will have specific bearing on the material well-being of the working majority in the advanced and emerging economies. By extension, the far-reaching discontent arising from this will have significant bearing on the hegemonic stability and capacity of capital globally.

The systemic capacity of labouring people to increase their consumption of capitalised goods structurally tends to plateau. This occurs, firstly, because every individual capitalist always tries to suppress labour's value share (in wages) of total production as a normal aspect of maximising profit. This of course has a cumulative effect. In the advanced economies wages have basically remained stagnant over the past 30 years. It is for this reason that Australia's Reserve Bank has extolled businesses to lift wages insofar as consumption rates significantly drop when overall wages are too low.

Secondly, consumption capacity diminishes systemically insofar as wages are a constantly reducing, variable component, of total capital employed in production. Capital constantly enhances the capacity of labour productivity by technologically revolutionising the means of production and thus increasing the proportion of up-front fixed capital costs expended relative to labour. Increasing the technological productivity of employed labour allows a reduction in the actual numerical and/or proportional cost of labour employed. This enables capital to generate and maximise profit in relation to its employed labour and its cost advantage in relation to capitalist competitors. The car industry deployment of robotics on the production line provides a case in point.

Productivity growth underpins a crucial aspect of the crises of realisation of the value (invested capital and profit) ingrained in capitalist commodity production. Capitalist growth, as an imperative, has nothing to do with altruistic dimensions for employing people, a living wage or improving their material quality of life. Capital actually structurally wants to do the opposite. Employ as few people as possible relative to what is invested to produce, pay as little as possible, and keep as much as it can for itself.

This of course diminishes the capacity of wage labour to expand their consumption of all those new commodities and realise the extra wealth they contain for capital's private appropriation and profit. Paying workers less ultimately means they buy less. But also problematic is over-producing commodities through ever-increasing productivity that exceeds effective demand. As we can see in the current Covid-19 crisis, this effective demand can fluctuate dramatically. Either you end up with more commodities than you can sell, or you end up with under-utilised capacity of capital invested in machines and raw materials. Not being able to sell produced commodities and get the invested capital back, including profit, is a first order conundrum for capital accumulation as a system.

Several important strategies have been employed to overcome this core systemic contradiction. Under globalisation, the costs of the material goods that reproduce the standard living conditions of wage labour (especially the core pillar cost of food) in the advanced economies have diminished, and their availability enhanced. In the advanced economies less than 20% of current income is spent on food (under 10% in America). One century ago it was close to 50%, as it is now in the less developed sectors of the global economy. Of course, this bald statistic masks considerable disparity across class structure, nationally and globally, including in regards the content and quality of the food available (consider here chicken McNuggets).

Increasing the scope and scale of production of primary commodities through capitalisation has, until now, been instrumental in suppressing wages. Cheapening the core costs of reproduction of wage labour creates space for the expansion of secondary commodity consumption with existing wages. The discretionary consumption of secondary commodities has also been substantially enhanced through cost reduction based upon economies of scale and re-locating production to lower waged regions. This has increased affluence (real and perceived) in the advanced economies in particular.

While the basket of goods that could be bought with largely stagnant existing wages has been significantly enhanced it hasn't been sufficient to mop up the total amount of commodities capable of being produced. Excess productive capacity and unsold commodities are a rolling feature of capitalist economy globally.

Simultaneously, expanding the neo-liberal privatisation and commodification of basic necessities such as water and other public sector assets and services, has extended the reach of capital into formerly sacrosanct areas. The search for profits knows few bounds and has been a major boon to those who have successfully established private ownership over what used to be considered public commons.

The massive extension and cheapening of consumer credit since globalisation have helped address this problem of accumulation, to the point now where most wage labour in the advanced economies are indebted for their entire working life to maintain their standard of living. This capacity for expanded consumption through consumer debt is reaching saturation point.

Also reaching saturation point is the capacity of ecosystems globally to absorb and regenerate in the midst of this breathtaking expanded production/consumption onslaught. Agricultural productivity has peaked, fresh water spoiled and diminished, soils degraded, fisheries collapsed, natural forests stripped, fossil fuels re-carbonised, atmosphere compromised, oceans acidified, heat intensified, and weather systems destabilised.

The gifts of nature have been inexorably eroded, largely beyond repair. These relatively free costs to capital for its continued expanded production will diminish and the absorbed fixed commodity production input costs will begin to escalate sharply. This is seen, for example, in the so-called extreme energy sectors such as shale-oil fracking and deep-sea oil drilling. So too, traditional forms of capitalist material production, reproduction and consumption will be radically disrupted as capital accelerates competitive technological innovation to address these costs dilemmas.

Mass migration of jobs and labour to low cost labour centres of production has been a key structural feature of globalisation. This may change with the deepening robotics revolution in the production sphere, leading not only to an accelerated mass reduction in employed production labour in the next 20 years but even the possible relocation of production sites back to the advanced economies.

Climate catastrophe will accelerate this mass dislocation stemming from globalised labour force restructuring. Migration from increasingly marginal, unsustainable and uninhabitable zones of the world economy is building and set to escalate. Rising global refugee numbers reflect concomitant rises in geopolitical tensions.

THE DEBT PROBLEM AND CLIMATE CRISIS

The consumer debt saturation enveloping wage labour in the advanced economies is being matched by a parallel debt saturation of corporate capital and the nation state. There will be tremendous systemic impact from the ever-greater proportion of infrastructure replacement arising from this multi-faceted global dislocation and the accelerating impacts of climate change. This will be a growing burden on both the social surplus generally, and national and international state budgets as a component of that. It will also add to the burden of labour reproduction costs, systemically and individually.

For productive commodity capital, extensive and accelerating revolutionising of the technological means of production – to pursue price advantage against competitors, minimise wages and maximise profit – requires ever greater up-front capital expenditure, and thus indebtedness, and ever-accelerating turnover times for capital realisation to recover investment. This puts individual corporate and system-wide downward pressure on profit margins over time. The size of the enterprise has to increase to offset this tendency toward relative decline in the rate of profit. The fragility of the enterprise, and the system as a whole, increases as even small interruptions in capital realisation become magnified. This has cascading and at times catastrophic systemic effects, whether it's due to superstorms, pandemics, or unaffordable 'liar loans' in the housing sector, as with the GFC.

For finance capital, finding ways to expand consumer, public and private capital investment debt is a raison d'être. Evermore exotic and perilous debt instruments are concocted to 'grow' finance capital and extract a rising usury share, aggregate and proportionally. The GFC and its aftermath remains a poignant and continuing illustration of its consequences.

State debt is structurally escalating from several sources:

- declining revenues from minimising the tax portion capital has historically paid, in order to prop up capital's declining profit margins;

- maintaining the social wage of 'welfare services' that labour struggles achieved as a result of capital's historical twentieth century expansionary period and the relative competitive advantage it enjoyed in the advanced economies;

- extraordinary money printing to re-inflate stagnating capital investment since the GFC, including and especially in the current Covid-19 crisis;

- State purchase of unviable and distressed private assets to avoid a chain reaction of financial collapse over unrecoverable debt.

None of this bodes well for capitalist expansion without substantial economic and geo-political crises and reconfiguration on a global scale. Much of this flows from the omnipotence of finance capital and the role of money in capitalist economy. It has multi-faceted dimensions, including a substantial illusory component.

Within finance capital's dominant centralisation and interpenetration in all aspects of global economy and social life are the seeds of its growing fragility and cascading consequences. At heart, it's a Ponzi scheme on a planetary scale. The proportion of debt increasingly bears no resemblance to the underlying value of global assets, the rate of return on those assets, or the actual currency in circulation.

No one in the corridors of power dares whisper the potential chain reaction of vulnerability of this fiscal confidence trick. Illusion and fragility in economic power also translates into the political sphere. Like all good Ponzi schemes, the authority of the minority rests upon the blinkered complicity of the majority. As we saw with the GFC it is a delicate balance that can and generally does unwind with astonishing speed and acrimony.

This is increasingly on display again as we lurch through the current coronavirus global pandemic. Post-GFC money printing has artificially inflated corporate profits and asset prices. Unlike the real consumer debt held by working people, this corporate and government debt operates in a parallel universe of 'too big to fail and too big to care'. Mega-corporate debt just sits there as basically 'interest free' money for corporate manoeuvrings such as buying back their shares, thus inflating the stock price and increasing executive bonuses.

For garden-variety corporates, debt is more directly connected with their need to expand productive capacity in order to increase corporate earnings. This has made the corporate sector ever more susceptible to interruptions in commodity-value realisation and thus to service even modestly priced debt, such is the scale of leverage. This degree of leverage is now set to dramatically unwind, as seen especially with the US shale oil industry. Massive cheap debt was a precursor to start comparatively expensive shale-oil production. That debt is now not capable of being serviced due to an unprecedented Covid-19 collapse in oil commodity prices. Corporate vessels (e.g. cruise companies and airlines) and stock markets are either sinking as a reflection of this ocean of debt or just barely staying afloat as the reality of pandemic-generated commercial paralysis has taken hold.

For too big to fail governments in the advanced economies that are operating reserve or 'safe haven' currencies (the US dollar), even more fantastical money printing at effectively zero interest is the 'operative' Modern Monetary Theory (MMT) style emergency crisis response. This theory holds that sovereign and especially global reserve currencies can't default because more money can just be printed as required, barring an outbreak of inflation of course. This hasn't happened...yet. But, for the bulk of globally indebted nation states, the 'kicking the fiscal debt can down the road' option either does not apply or not as much.

Amidst this growing interconnectedness and fragility, the climate crisis and the transformations required in the face of it, are a toxic political brew for continued capital growth. This is now openly being recognised within the corridors of power itself. The debt problem has been a component element of deliberate climate denial after all, making continued rational debate with vested climate deniers even more absurd 40 years down the track. It is increasingly splintering the political wall of silence of those seeking to manage threats to the 'collective interests' of this cabal of competitive self-interests ruling over our future well-being.

To illustrate this, consider recent headlines and reporting on the cracking façade within the corridors of power:

Richard Partington, 'Bank of England boss says global finance is funding 4C temperature rise: Mark Carney says capital markets are financing projects likely to fuel a catastrophic rise in global heating', *Guardian*, 16 October 2019:

> ...in a stark illustration of the scale of the decarbonisation challenge facing the world economy, Carney suggested companies had already secured financing from investors in the global capital markets – worth $85tn for stocks and $100tn for bonds – that will keep the world on a trajectory consistent with catastrophic global heating.

Larry Elliott, 'World economy is sleepwalking into a new financial crisis, warns Mervyn King: Past crashes spawned new thinking and reform but nothing has changed since 2008 banking meltdown, says former Bank of England boss', *Guardian*, 20 October 2019.

Phillip Inman, 'IMF boss says global economy risks return of Great Depression. Kristalina Georgieva compares today with "roaring 1920s" and criticises UK wealth gap', *Guardian*, 18 January 2020

> ...the global economy risks a return of the Great Depression, driven by inequality and financial sector instability...She warned that fresh issues such as the climate emergency and increased trade protectionism meant the next 10 years were likely to be characterised by social unrest and financial market volatility.

In reporting the summary conclusions from the 'green swan' report (which relates to climate change risk factors) by BIS (the Bank for International Settlements – 'the Reserve Bank for Reserve Banks'), Stephen Letts, 'Reserve Bank urged to battle "green swan" risks of climate change', *Australian Broadcasting Corporation (ABC) News*, 22 January 2020:

> The BIS Green Swan report argued current risk assessment and climate change models cannot anticipate accurately enough the form that climate-related risks will take and the 'potentially

extremely financially disruptive events that could be behind the next systemic financial crisis.' The value of the world's stranded fossil fuel reserves is massive but difficult to pin down according to the report, ranging from Carbon Tracker's $US1.6 trillion ($2.3 trillion) esti-mate to the International Renewable Energy Agency's 2017 calcu-lation of $US18 trillion ($26 trillion).

and

Due to investments in and loans to fossil fuel producers, financial risks would spread throughout global markets leading to a mass of defaults and tightening banks' liquidity and ability to lend, in much the same way as the contagion caused by the collapse of the US subprime mortgage sector spread rapidly worldwide during the 2008 financial crisis.

More recent reporting on the start of the pandemic economic crisis reflects the expansion and scale of the debt strategy response by central banks globally, despite the already staggering levels of debt-fuelled expansion going into this recent economic crisis.

Julia Horowitz, 'Cash handouts are coming as countries do "whatever it takes" to survive the pandemic shock', *CNN*, 19 March 2020:

Borrowings from households, governments and companies grew to $253 trillion during the third quarter of last year, according to the Institute of International Finance. That put the global debt-to-GDP ratio at over 322%, the highest level on record.

In these articles leading banking and financial figures – such as the head of the IMF and the Bank of International Settlements – warn that issues such as climate change and increased trade protectionism will be accompanied by financial market volatility, global financial crisis and social unrest. This includes trillions of dollars lost due to the stranding of fossil fuel assets worldwide, rapidly escalating world debt, and mass defaults on bank loans. This from the horse's mouth.

Of course, these trends will only soar with the economic crisis ramifica-tions of the coronavirus pandemic. Moreover, the pre-conditions for further cataclysmic national and global climate economic impacts are already in place for when the current pandemic emergency passes.

The debt problem is also a toxic political brew for our movement for ecological and climate action if we fail to seize the moment to change course before unpalatable changes are foisted upon us – financially and politically. In this regard, a green capitalism is as much an oxymoron as 'no-growth' capitalism.

The Green New Deal (GND) in the American progressive movement is predicated upon the same Modern Monetary Theory (MMT) that has been employed in practice since the GFC and in the current post neo-liberal budget deficits of the Trump administration. Giving it a pre neo-liberal, Rooseveltian political reform wash has provided some supportable momentum in the United States – particularly given its coupling with social welfare measures like universal public health care that is taken for granted in other advanced economies. But basically, it's a 'tax and regulate the capitalist ruling class' strategy. Nothing more, and possibly a great deal less if history is anything to go by.

Mimicking this GND policy language by the green movement in other advanced economies, such as Australia, doesn't just lack originality by importing a historical reference that only really culturally resonates in the American context. It also seeks to draw limited social democratic parameters around the change being advocated in this climate and ecological emergency. It was war, not Roosevelt and the New Deal, that ended the Great Depression. The New Deal was a hiatus in the full throttle of capitalist expansion because the engine itself was misfiring and needed maintenance. It didn't replace the engine.

Until the levers of private versus collective ownership of the means of production and socially produced capital are directly addressed then full throttle private wealth expansion will resume sooner or later. This developmental contradiction lies at the heart of the CMP and is becoming more and more transparent. In a nutshell, the increasingly *interdependent and interconnected nature of the forces of social production and the climate and ecological emergency* is increasingly at odds with the *concentration and centralisation of private ownership*, encompassing the appropriation and accumulation process and key investment decisions.

For all the apparent omnipotence of the ruling 1% and the mega-corporations, power and control has its limits as well. The structural antecedents for resolving this systemic contradiction in the direction of a post-capitalist future are becoming more pronounced. They are just as historically compelling as how the growth of money capital and its possessing class created the historical demise of feudalism and absolute monarchy centuries ago. We address this more directly in later chapters.

The post-Holocene transition of unsustainable endless growth amid ecological collapse is also a transition of recurring economic dislocation and crises, and of increasing socio-economic marginalisation and political discontent. This is taken up in the next chapter.

4

FRACTURING CONSENT: MINIONS, MERCENARIES, MALCONTENTS AND LES MISÉRABLES

Capitalism has proven to be remarkably resilient. Since its full trans-formation into generalised commodity production there have been countless cyclical crises of accumulation of varying degrees of severity. For the 1% ruling class that own over 50% of the wealth these crises can be both inconvenient and opportune. It depends upon the competitive circum-stances. For the balance of our class-divided societies however, capitalist crises come with considerable devastation for our present material cir-cumstances and future prospects.

As we have entered, yet again, an economic crisis of still uncertain but considerable magnitude, it begs the question of why an economic system built to primarily and unequally advantage such a tiny minority remains so resilient?

The answer to this question falls within the domain of capitalist hege-mony. The ruling economic, political and cultural dominance of minority capital ownership and control over the interests of a dependent majority of subordinate classes has to be perpetuated over time. To be sure, it requires the capacity to exercise power directly (for example, through coercive instru-ments such as the military and police enforcing legal-juridical property rights). But it also requires conceded legitimacy and consent on the part of the dominated majority.

These direct and 'soft power' processes ensure an embeddedness of compatible cultural givens. For example, the 'self-evident' and extolled virtues of 'private property', individual responsibility for getting ahead, existing institutional arrangements of social order (including corporations), and taken-for-granted class inequalities, are ingrained in everyday life and 'common sense' thinking.

This is not to say that this hegemony over subordinate classes is uncontested. On the contrary, capitalist hegemony has never been complete. The contradictions and unequal consequences that underpin recurring economic crises cannot be papered over and ignored.

Like the Romans fretting over slave revolts, the democratic repercussions of the current Covid-19 pandemic, for instance, are fundamentally shaping the emergency budgetary response of capitalist governments globally. Neo-liberals have converted to Keynesians overnight. Similar to the Rooseveltian New Deal in Depression-era America, 'debt and deficit' is now being embraced to maintain the *consent conditions* for restarting 'business as usual' accumulation when the economic health emergency passes. But, as with the climate emergency that is still upon us, the political contest for the shape of this systemic restart still lies ahead.

In this chapter we discuss the configuration of the class structure of the global capitalist order, the cross-class hegemonic reproduction that capital requires to rule, the fracturing of the cross-class consent required to contest that rule, and the extent to which that hegemonic fracturing is currently underway.

Escalating ecological crises dovetail with rolling accumulation crises to create counter-hegemonic options and opportunities for building a transformational movement to address the climate and economic emergency. In order to succeed we need to directly address the structural underpinnings of ruling class power and interests that systemically create these emergencies.

WHY THIS MATTERS

Class structure is important as it embodies the core foundation of social relations and material interests that shape all other societal relations and identifications. Economic power or lack thereof is the bedrock of social and political power. Any movement for economic, social or political transformation that threatens ruling class interests and power has to build an alternative power base to counter the inevitable and myriad forms of assertion of core dominant interests. This has been transparently obvious throughout the failing 40-year struggle to avert the climate and ecological emergency that is now upon us.

Building that alternative power base *requires* galvanising majority subordinate class interests. The most immediate obstacle to achieving this is that subordinate class interests are dependent upon dominant class interests, even if there is a core structural antagonism and relation of exploitation as it is with the capital-labour relation. Subsidising technically insolvent employers so that they can continue to employ their workers in the current crisis reflects and captures this required dependency relation.

An abstract objective political interest to overturn the capitalist relation of exploitation on the part of labour (making profits for owners) doesn't automatically translate into political indifference on the part of working people to their immediate material interest in maintaining the employment necessary to pay rent, buy food, cover debts and secure family well-being.

In conditions of capitalist crisis, systemically conflicting class interests are exposed by the extent to which the dominant class is not able to deliver sufficiently on the immediate dependency requirements of subordinate interests. This creates the political conditions where abstract objective interests and immediate material interests converge for subordinate classes.

Even then, in that convergent political moment where the 'legitimacy/ acceptance' of the dominant-subordinate class relation may fracture, an integrative counter-hegemonic narrative is still required. Subordinate class interests have to be subjectively galvanised into a transformational political force. Economic hardship alone is never enough to generate a transformational, counter-hegemonic direction.

This counter-narrative absence was evident during the GFC (the Occupy Movement notwithstanding), is still absent in the global economic crisis unfolding in the current coronavirus pandemic, and in our continuing lack of political cut-through in the climate emergency.

For example, in the last Australian election cycle (2019), the *Stop Adani* movement (against a proposed mega-coal mining project) organised a sustained march into the drought-stricken and economically crippled heartland of regional Queensland. Wide political support against further coal mining exists within urban population centres nationally. Importantly, years of economic failure of conservative political representation regionally as well as nationally has seen political fracturing in regional areas. The continued legitimacy and political allegiance to those who have politically overseen the integrated regional economic crises, and the corporate agricultural and mining interests they serve, has been under considerable strain and challenge. The political situation is fluid, with the fracturing being in both (mostly) 'right' and (some) 'left' directions. More accurate though would be to say that the fracturing is *politically unformed*, being simultaneously very real but directionless.

It's difficult to fully comprehend the desperation of these drought-stricken regional areas in question unless you live there or have seen it up close. They are truly at the front line of climate change in Australia, along with regional communities in the recent bushfire zones. Small intergenerational family farms, most already economically marginal from the competitive onslaught of corporate agriculture and the food retail giants, are collapsing. Stock losses are widespread, through forced euthanising or sale at basement prices. Water is being trucked in on a massive scale, just to cover minimum essential personal use. There are no jobs for young people, who are migrating to the cities for work. Small businesses and social/health services in local towns are shutting up shop on a wide scale. Suicide rates are the highest in the country.

The political effect of the regionally targeted *Stop Adani* march was a visceral reaction to inner-city and inter-state outsiders coming in to tell them what they could or couldn't do to deal within this existential crisis. It drove vacillating regional voters back into the very arms of the corporate interests (and their political minions) who have been overseeing this regional calamity. The unexpected neo-liberal government re-election has now led to removing existing political logjams that had already been created by our opposition movement to the mine proceeding. As part of the post-election repercussions, soft political support for a post-coal transition policy agenda from less dinosaur sections of the main Liberal/Labor ruling parties, is now crumbling. We need to learn from this.

It's not enough as a movement to demonstrate that we think climate change is an existential crisis that others should care about. And it's partic-ularly not good enough to march in and demand climate emergency front-liners 'take one for the team' and reject a desperately needed, 'new job-promising' coal mine as if it's a half-time pep talk in a losing football match. Their crisis is as materially immediate as it is existential in the longer term. We need to build and sell an immediate and secure long-term economically sustainable transition plan for those already in climate crisis. Otherwise we will continue to fail in galvanising the breakthrough democratic support required to make the difficult structural transformations for ecological survival. Even the United Nations Environment Programme has enunciated the need for a policy framework for transitioning away from coal, including regional and economic diversification strategies.

Leaving vulnerable communities in the lurch is not good political strategy. Nor is simply prioritising our convictions of the necessary response to the climate emergency over the emergency versions being confronted by others. Even if we consider extraction industry communities to be 'hostile' political territory, with majority material interests contrary to ours, they still contain a subordinate class support base we need to win and cross-class ruling interests we need to neutralise if not fracture.

Frankly, most of regional Australia already get that the climate crisis is real and are even amenable to acknowledging what the origins are, especially in the rural sector. Ruling corporate interests (and through their State-based minions and mercenaries) have the resources to dangle just enough carrot to feed self-interested denial and/or paper over the class-alliance cracks. It's the same formula used with just drip-feeding the prospect of a greener capitalism in general that keeps our movement too unformed, contained and directionless.

Ruling interests have the social/political strength that flows from economic control of systemic subordinate class dependency. Unless we reach out, integrate and resolve those immediate subordinate class crises as part of the overall climate emergency then our movement will continue to get politically outflanked and overrun by dominant class interests.

CROSS-CLASS CONSENT

Throughout the history of formation of complex societies and economic organisation in the Holocene there has always been some form of class structure that emerged to produce and appropriate the social surplus. There has also been an institutionalised hegemonic capacity on the part of the dominant minority class to maintain its privileged position. Under feudalism, for example, this was expressed in the doctrine of the 'divine right of Kings and Queens' to hereditary land ownership and rule. Custom and tradition reinforced the everyday dominance of this minority, as did the class layers of land-holding lords, their standing militaries and organised State religion.

However, class structure and hegemony under capitalism is historically unique. First, possession or share of the social surplus is not defined by class position but rather class position is defined by possession (or not) of the means through which the social surplus is produced. Second, the class relation for how that social surplus is produced and appropriated is defined economically rather than by extra-economic criteria such as whether one has (or not) 'divine right' by birth, political or military power, or assigned/ institutionalised cultural authority. Third, the exploitative economic relation between the class that appropriates and the class that produces this social surplus is structurally antagonistic by definition.

As established in Chapter 3, the core economic class relation that defines capitalism (the shape of possession and its inherent antagonistic character) is the relation between those who own capital and those who only possess their own labour power. Possessors of capital live at the expense of those who have to labour to live. Those who only possess their labour power need to sell themselves to those able and willing to buy it. This identifies the inherently antagonistic class relation of exploitation between capital and labour that lies at the core of capitalist economy.

How that core abstract relation plays out in concrete social life in capitalist societies is of course multi-faceted. In the advanced economies in particular it is often lived experientially as a contradictory set of relations – where say, through one's pension fund capital investments or an investment property, you can be both a worker and an owner of capital. This often has a lived material ideological effect on how you see yourself within a class-divided society, if not fundamentally shaping your view of the society itself.

Class is often refracted and obscured by ideological, cultural and political factors. Pervasive is the everyday political/cultural language used to discuss class in distributional terms – as in how much money you have or education or managerial-type functions at work. These are of course real enough lived experiences (and shape comparative self-identifications). But it doesn't tap down to the political and explanatory core of what shapes social class structure within a capitalist society and, more importantly, how that class structure shapes contemporary patterns of hegemony.

For example, as far as distributional markers go, there are Fly-in/Fly-out (FIFO) remote mining workers in Australia who make triple the income of small retail business owners and more than many in the 'white collar' and 'professionalised' sectors. Many people surveyed (even the academics doing the surveying) in the advanced economies think that there is a 'middle class' that starts at US\$50,000–60,000 a year (household). Labour force restructuring under neo-liberalism, and more recently with the 'Gig' economy, turned 'employed wage workers' into structurally and ideologically re-classified self-employed 'small business owners', 'sole traders' and 'independent contractors' overnight – suppliers of their own tools/means of production, 'business loan' capital, sick leave, and retirement incomes. Ideological and political spin and news fakery didn't just emerge with the ascent of a barking orange clown to the Imperialist throne in Washington.

Even within our own progressive movement, and particularly since the post-GFC Occupy Movement, we loosely talk of the 'them' 1% versus 'we' 99% and even the 99% economy. In terms of isolating politically and ideologically on the absurdly small numerical minority that is the core ruling class within global capitalism, that's fair enough, even generous. But as an explanatory vehicle for the social class architecture that maintains hegemony, and how it could fracture and dynamically overturn the course of our mode of self-destruction, not so much. 'We 99%' are not 'all in this together'.

Regardless of the swirling spin and self-identifications, the structural core class relation of capital versus labour must always be preserved and is never far from the surface of how social life carries on. Often it is just subconsciously accepted and explained in non-pejorative terms as the way life is, always has been, and can only ever be. Generating and preserving that condition of acceptance of class exploitation (even a grudging rather than unconscious acceptance or even as aspirational class mobility – such as relatively well-paid workers borrowing money capital for investment properties) is the domain of capitalist hegemony.

Capitalist hegemony is distinguished historically by virtue of the political rule of the dominant class being an indirect rather than direct form, function and expression of institutionalised power.

Capitalist class structure and hegemony consolidates as private ownership and control of the means of production and appropriation of capital, and a companion State political and legal/juridical institutional apparatus that secures its functional stability and authority, especially that of the sanctity of private property. Of course, there is also a repressive apparatus to back it all up, just in case. In this sense, the parallels with previous forms of economic organisation and institutionalised hegemony in class-divided societies are evident. But there is one crucial difference here as well. As with economic dominance, the apparent hegemonic omnipotence of institutionalised capitalist power also contains its own unique seeds of fragility and collapse.

The ruling class within the CMP (like all ruling classes) rules through economic power, repression, and sufficient consent from subordinate classes. For the 1%, capitalism today requires cross-class consent to rule on a scope and scale qualitatively unlike any previous ruling class in history. The bourgeois democratic revolutions of the eighteenth century, while significant in overthrowing feudal relations and monarchical rule, did not fundamentally change that equation from pre-capitalist periods. There were two significant core components that shaped that transition:

- Firstly, the extension of private property ownership became the principal basis for social class structure. This sub-divided into ownership/control of increasingly centralised means of commodity production, distribution, exchange and finance, on the one hand, and the small-scale propertied (partially labouring) petty-bourgeoisie on the other.

- Secondly, the State apparatus became predicated upon this multi-layered propertied class structure, both in terms of economic interest management and political legitimacy/control.

There was initially only greater diffusion of institutional political control to incorporate democratic voting rights for subordinate *property-owning classes*. In some countries (France) heads had to roll as part of the ruling class transition. In others (England) gentlemen's agreements settled the expanded ruling class transition.

The subsequent extension of democratic voting rights to the non-propertied working-class majority (starting with men some 100 years later and then women at the turn of the twentieth century) has undeniably changed the political dynamic of class rule. It needs to be noted, of course, that this democratic extension to the non-propertied majority was only because of the latter's 150-year struggle for universal suffrage against the expanded sole rule of propertied classes. It was not a part of the 'bourgeois democratic revolutions' at the time, however much it is claimed as part of the historical mythologising of liberal democracy in the 'Western' advanced economies.

Even today, in that pantheon of liberal democracy that is the United States of America, there is currently an active push in the Republican-run swing States to disenfranchise hundreds of thousands before the 2020 presidential election. Similar occurred in Florida with thousands of (mostly black) voters with 'criminal' records disenfranchised during Bush versus Gore. All this in a democratic exemplar where already barely more than 50% of eligible voters actually do so in any given presidential election. In Australia, which has a relatively unique electoral requirement that involves compulsory voting, removing this compulsory component has been part of the political agenda of conservative parties for many decades.

Over the past century of contemporary capitalism, cross-class consent has come to play a much larger part of the equation to consolidate and maintain hegemonic control. We also need to be clear what 'consent' consists of. It is certainly a grudgingly conceded consent on the part of the dominant class. As noted above, it doesn't stop machinations to restrict, divert or ignore democratic franchise or even results. For example, even democratic minimalism is presently under threat in many parts of the world (e.g. the Philippines and Russia) and currently being undermined by circumstance and opportunity (e.g. the new absolute powers of the President of Hungary in response to the Covid-19 pandemic).

For many in subordinate classes, consent can be as minimalist as just not rebelling, at least directly. This is understandable given the powerlessness and precarious position most find themselves in at an individual level. Mostly though, structural cross-class consent has flowed from the expansion of the propertied middle strata in advanced sectors of contemporary economy and thus as a proportion of the total class structure. Consent has also flowed from some additional sharing of private material wealth accumulation across working class layers (e.g. more recently reflected in privatised retirement income within investment funds to accompany the existing 'social wage' of State welfare services). Along with the enhanced 'basket of goods' still accessible despite stagnant wages it has been an affluence-based consent and part of the commodified accelerant underpinning the ecological extinction pace.

Of particular note is that the expansion of the cross-class structure of capitalist rule is embodied within an expanding 'relative autonomy' of the State apparatus. The State structurally integrates and incorporates subordinate class interests in political and ideological terms. It is the institutional lynchpin for capitalist hegemony.

It is important to clarify what this 'relative autonomy' consists of. It is *not* autonomy (relative or otherwise) *from* ruling class interests. It is a Capitalist State that *directly* upholds and reflects the general and specific interests of Capital. It adjudicates the competing interests of capitalist sectors where required, and provides the political, legal-juridical, financial, administrative, repressive and infrastructure functions required to manage and grow the capitalist economy and solve its periodic accumulation crises.

The 'relative autonomy' reflects/flows partly from the sheer expanding scale, scope and complexity of the State apparatus in accordance with the expansion of the national and global economy. There is a historically evolved and primary economic centrality of the State. The State apparatus is a necessary, functional and directly integrated operational entity of capitalist economy in itself. Moreover, it directly employs a substantial proportion of the labour force across the advanced economies (in the order of 15–20% depending upon how it is counted), plus being a substantial contractor of middle strata services in areas such as technology, finance, policy research and marketing. Relative autonomy also flows from the 'consent requirements' to integrate the interests of subordinate class sectors and the added functions that flow from that – especially the social services to manage the labouring classes for Capital and to massage the precarious systemic dependency and advantages of subordinate propertied middle strata.

These clarifications of relative autonomy are important because they contrast with several strategic political alternatives that have dominated our broad left movement over the past century. The traditional Leninist/ revolutionary socialist formulation on the need to *smash the capitalist state* and create a new socialist workers state has become quite anachronistic after another century of economic development. Even allowing for a considerable degree of rhetorical political flourish, it is a formulation that reflects more historically embryonic and transitional capitalist state formations that still incorporated pre-capitalist (Tsarist) nation-state forms.

The capitalist state would indeed be fundamentally transformed in a revolutionary post-capitalist fashion. But this will need to be done upon the basis that the development of the contemporary Capitalist State already reflects and embodies antecedent structural components, evident within the wider structure of capitalist economy, for the future eco-socialist economy and its structurally democratised State form.

More pressing is the need to clarify 'relative autonomy' in relation to the social democratic version of the Capitalist State that governs the bulk of our left movement political strategies, particularly that which underpins the Green New Deal and its derivative versions. Here the view of relative autonomy is predicated upon the formulation that there is partial or considerable autonomy of the State *from* the interests of the ruling class and the systemic parameters of the economy. Implied here is the idea that the State can, through elections, become 'our' State not 'theirs'. However, today this typically is not even framed in terms of the 'parliamentary road to socialism' from previous movement debates. Rather, it is conjuncture-tied to election cycles and 'green capitalist' budgetary and social policy priorities. It is not viewed as an integral component of a wider, post-capitalist transformational project.

The foundation of this reform agenda is that the State has an existing independent power that can allow it to control, regulate, tax and re-direct *economic management* regardless of the core interests of capital and the systemic demands of growth and accumulation. This conceptualisation of the State thus doesn't require expropriation of the central levers of ownership and control of the means of production (hence omission of talk about the overall socialist mission). For instance, nationalisation of some core areas of economy (such as banks or airlines) is less 'transitional' or an 'interim step' to a new mode of production, and more of a 'shield' or 'shock absorber' for subordinate class interests to survive against the ravages of dominant class interests and the systemic crises they engender. The core dominant interests and systemic imperatives are thus preserved, not contested, even in emergency conditions of crisis.

The most exceptional form of the emergency capitalist state historically was the Fascist State in Germany. The apparent political 'independence' of the fascist state in the emergency conditions of capitalist crisis was actually one of direct structural fusion with the dominant interests of German capital and the direct political mobilisation and integration of subordinate class interests into that common national capitalist project. The Chinese Capitalist State today shares some of these features, but in a fusion that foreshadows a potential transformation into a permanent new form of State Capitalism globally.

For liberal/social democratic versions of the emergency Capitalist State in conditions of crisis, the exemplars are found in times of war and depression. For instance, the Rooseveltian underpinnings of the GND apply here, as do the enhanced authoritarian possibilities emerging in the context of the current Covid-19 global pandemic. The apparent independence/autonomy of the State to marshal the central levers of the economy without expropriation obscures the actual direct political fusion *with* the dominant ruling interests in conditions of emergency. Without expropriation it becomes a more direct political fusion based upon *less democracy* and suspended active cross-class consent, not more.

As the climate emergency deepens so too will the emergency conditions for liberal capitalist government. It follows that the democratic components of liberal democratic capitalism will be increasingly constrained by the systemic demands of the authoritarian private ownership and control structures that the State is grafted onto.

Voting every three or four years for cross-class representatives to occupy the Capitalist State apparatus and carry out overseer functions for both ruling and subordinate interests is a long way from the structurally encompassing democratic socialist forms that are required for realising subordinate class interests in conditions of crisis, and for comprehensive transformation. But it has been enough to secure ideological consent of subordinate property-owning and materially advantaged class sectors in normal circumstances. It's also been enough to politically secure the operational 'relative autonomy' of a State apparatus to maintain cross-class capitalist business as usual. Why that cross-class consent has been so stable, and under what circumstances it may fracture, have to be clarified.

FRIENDS WITH BENEFITS: THE PETTY-BOURGEOIS MIDDLE LAYERS

The capital-wage labour relation lies at the core of capitalist economy and society. Nonetheless, capitalist hegemony, and the increasingly socialised character (scope and scale) of the forces of production, demands and generates intermediate middle strata that are integrated within its hegemonic structure.

Relative global advantage in capital growth and affluence in the advanced economies, over the post-war period especially, has generated a wide layer of intermediate strata of minions and mercenaries required to carry out that expansion and enabled sharing in the accumulation spoils. These middle strata occupy strategic structural and political weight within contemporary economy and the State apparatus. For example:

- Middle managers, stockbrokers, software engineers, investment and marketing advisors, economists and accountants permeate the corporate sector.

- Bureaucrats, lawyers, academic research, economists, communication and logistics experts likewise inhabit the State apparatus, beyond the regular administrative and policy staff. Then there is the entire legal-juridical and repressive apparatus of the standing militaries, police, border control, intelligence and so forth.

- Small-scale petty-bourgeois enterprises, from doctors and dentists to contracted professionalised business services (e.g. accountants, marketing, labour hire) to corporations and the State, coexist with traditional shop-keepers and family farms.

There are important multiple dimensions that distinguish these layers of middle strata from Capital, Labour and each other. But what unites them is this: all middle strata embody (if only abstractly) features of both capital and labour. All have ownership or control functions in the economy or the institutional apparatus of the State or civil society, but all have to labour themselves at the same time. All are dependent upon capitalist hegemony economically and institutionally, and all are insecurely subordinate to Capital's growth requirements and systemic crisis dynamics.

The propertied and control function component of middle strata marks their systemic political interests and integration with the interests of the ruling class. Their dependency in relation to Capital is different from the dependency relation of labour per se. The latter dependency rests solely and fundamentally about the requirement to sell their labour to live, irrespective of the level of remuneration – even though relative income advantage may provide a buffering of sorts for some.

For middle strata, capitalised property ownership and/or delegated managerial control functions materially and directly connects their dependency upon the reproduction of the economic system itself. Whether as traditional property-owning shopkeepers or as better remunerated professional and managerial layers with investment properties and share portfolios, dependency also translates into abstract and concretely shared interests. While less than 1% of the population own over 50% of total social wealth, the top 20% collectively own over 80%.

The expansion of middle strata as an integrated and necessary component of growth and accumulation in contemporary economy (whether as small business contracting, or as managerial functions in the corporate or State apparatus) creates a structural cross-class consent. Moreover, it embodies, by extension, a structural antagonism to employed labour – at least abstractly. It is a conservatising structural and political/ideological location.

It also underscores a particular middle strata relation to the Capitalist State. The subordinate relation to dominant corporate interests, and the dispensable competitive commercial position of middle strata to systemic dynamics of accumulation and crisis, creates structural insecurity. Moreover, the structural obligation for middle strata to also labour themselves creates a comparable structural precariousness to that of labour generally. The underlying fear of being more directly cast into the ranks of the labouring class is ever-present. The cross-class relative autonomy of the nation-state is the only political leverage middle strata have as a shield against both corporate capital and organised labour, particularly in periods of relative crisis. They see, expect, declare and demand the Capitalist Nation State to be 'their State'.

FRACTURING CONSENT: MALCONTENTS AND LES MISÉRABLES

Capital-centred forms of institutional State power are a principal obstacle for any transition to new relations of economy and ecology. By extension, far-reaching discontent within the heart of middle strata will have significant bearing on the hegemonic stability and capacity of Capital globally.

Proletarianisation under 40 years of neo-liberalism has pushed greater numbers of employed institutional middle layers (within the State apparatus, corporate structures and civil society) into more precarious conditions comparable to that of advantaged sectors of the working class. In essence, there has been a decline of comparative advantage in income and job security over that of domestic working-class sectors.

Similarly, the relative advantage of domestic working-class sectors over that of the working-class layers in the less developed sectors of the global economy has also declined with globalisation of manufacturing, technology and financing. Wages have stagnated and industries have relocated. The dramatic workforce re-structuring stemming from information technology and robotics that marks the third industrial revolution looms on the immediate horizon for all employed middle-strata and labouring sectors.

Climate change events, both as immediate perils and in the form of forced longer-term adaptations and mitigations, are exacerbating the more precarious conditions emerging economically for subordinate classes. The rich are insulated, particularly in their capacity to recover and adapt. Bill Gates apparently has bunkers in each of his six houses and New Zealand has seen a boom in the inflow of billionaire capital to secure a survivalist bolthole if worst case climate change scenarios unfold. But vast layers of traditional property-owning middle layers are not insulated. Droughts, fires, storms and floods don't discriminate according to class. They decimate farms (large and small) and urban and regional community shopkeepers and small businesses just as much as the workers they employ and the homes in which they live.

Growing discontent with globalisation across all advantaged layers of middle- and working-class strata is currently being reflected politically and that will escalate. There was a fracturing of the liberal/social democratic hegemonic centre underway prior to the current pandemic. Trump, Brexit and the surge of the Ultra-Right across the advanced economies, particularly Europe, reflects that a dramatic re-shaping of the political landscape is underway. This will likely exacerbate as the convergent consequences of the current accumulation crisis and growth restructuring within an ocean of debt fully unfold. Restructuring tensions will be further heightened by attempts to withdraw financial stimulus measures introduced with the coronavirus pandemic, and as climate change economic and ecological effects deepen.

Manufacturing ongoing consent in the advanced capitalist economies has a number of components. The strategic role of social democratic measures to buffer the negative cyclical material effects of capitalist accumulation on the working classes through greater social distribution of wealth is one factor. Also critical has been the entire architecture of the ideological and cultural apparatus of the Capitalist State and its shaping influence throughout 'civil society'.

Seemingly, only the masters of our capitalist universe and their spin merchants now talk about the perilous fragility of capitalism. Of course, for them it is the need to actively save it, like during the GFC and what is to come from their current Keynesian deviation caused by the Covid-19 pandemic. They seem more acutely aware of capitalism's contingent character than we are. To the rest of us capitalism, its growth protocol and class structure, seemingly presents as an omnipotent juggernaut incapable of being over-thrown, transformed, challenged, or even questioned.

Our everyday lives are so embedded within a pervasive global capitalist social order that what is a historically specific and evolved set of social and economic relations has become normalised as eternal and axiomatic in popular consciousness and culture. That's an important part of the equation of why political 'cut through' for structural transformation has eluded both the climate action movement and anti-capitalist movement generally.

Green capitalism represents the reform limit of most climate movement political agendas because capitalism just presents and gets presented as 'economy', as an eternal ahistorical given.

The politics of greening capitalism parallels the history of 'equality of opportunity' politics. The culturally lived 'material ideological effects' of capitalist hegemony are more pervasive than cross-class group identifications and politics superimposed upon or integrated with class structure (such as gender or race/ethnicity and the struggle for 'equality of opportunity' to be evenly represented/distributed across the unequal class structure). Just like with the women's movement and the black rights movement, 'greening capitalism' will not automatically translate into the ecological equivalent of the equality of condition transformations that many were purportedly intending to seek. Eco-capitalism will still give us a capitalist ecology for profit generation. The experience in other social movements shows that initial reform momentum is typically re-directed toward narrower interpretations of 'equal opportunity', such as having greater chances to be integrated into the advantaged layers of the structure of power. This is the essence of the 'glass ceiling' narrative in regards women.

Not only did this significantly de-mobilise these 'Rights' movements back in the 1960s through to the 1990s, the initial concrete objectives still continue to elude the very 'equal opportunity' agenda the left social democracy wings of these movements claimed as achievable. Yes, there have been gains, both for minority rights and in greener climate action transitions. More women and racial/ethnic minorities sit in corporate boardrooms. Some (like Thatcher and Obama) became heads of government. Distinct strategies to build black and multicultural 'middle classes' across the advanced economies and State apparatus have been realised, and there are now considerable middle strata career paths for helping to green capitalism across the corporate and State sectors.

But most working-class women and racial/ethnic minorities are still waiting for the claimed cross-class political benefits of equal opportunity and equitable treatment. The George Floyd/Black Lives Matter demonstrations in particular show just how limited equal opportunity transformations have been in the working-class lives of the 20% African heritage minority that comprise contemporary America. This, some 60 years after multiple Civil Rights Acts. In Australia, Indigenous deaths in custody have averaged in excess of one per month over the past 30 years. This, *after* the landmark Royal Commission into Aboriginal Deaths in Custody report. Indigenous Australians constitute just 3% of the adult population nationally but over 28% of the prison population. If this isn't startling enough, Aboriginal children are 4–5% of the national youth population yet make up over 50% of those incarcerated in juvenile detention centres.

These examples, as with the spontaneous global support to the American 'Black Lives Matter' movement across the advanced economies, *are not a reflection of the unfulfilled promise* of liberal capitalist democracy. They reflect the cross-class consent success of liberal capitalism's hegemonic parameters and its abject democratic failures for subordinated classes, social groups and the ecological conditions of social life for the national and global majority.

The climate emergency doesn't have 60 years to achieve partial re-styling of the core business landscape. More to the point, movement strategies seeking to moderate ecological destruction, as with the equal opportunity movement seeking to moderate inequality of subordinate class condition, have become subsumed as a component within the very definitional framing, acceptance and ultimate reproduction of the system itself.

Cross-class identities are not a 'false consciousness' or distraction central to maintaining the cross-class hegemony of Capital. That old left formulation was never theoretically valid and, to the extent it still survives as a political characterisation, has long needed to have been set aside. The structural power and resilience of capitalism and its ruling class fundamentally rests upon its institutionalised interests being shared by the dependent, immediate material interests of subordinate property-owning class sectors. This extends to the immediate material dependency of working people, who only survive through sale of their labour, and thus have a shared immediate material interest in the viability and survivability of the employing enterprise.

Hegemony rests upon capitalism's capacity to deliver upon these imme-diate material interests for middle strata and working-class sectors. The need to ideologically and politically develop a class consciousness among subor-dinate classes – that their structural interests extend beyond these immediate dependent material interests and that these broader interests are in structural conflict and contradiction with those of the ruling class – does not mean that there is some 'true' level to which that self-awareness has to develop. Nor does it mean that the opposite 'falseness' applies if that extended class interest awareness has yet to emerge or is not sufficiently formed to decisively act upon.

Cross-class identities are an integral component, rather than a decisive obstacle, to fracture and contest capitalist hegemony. In the case of the oppression of women and racial/ethnic minorities, cross-class identities are an important and central component of overall lived class experience through which class identities are mediated. The historical use of racism and sexism to shape class condition and division within capitalist societies, as well as cross-class re-calibrations of hegemony (as with 'equality of opportunity' and 'meritocracy'), are part of how ruling class hegemony is institutionally reproduced over time. These class integrated identities are thus a crucial ideological and political component for fracturing ruling class hegemony rather than reproducing that hegemony.

Patriarchy, colonialism and slavery have enduring cultural resonances (the habitus of 'whiteness'), institutional legacies (evident in income and employ-ment gaps, amongst other measures) and interpersonal expressions (mistrust, superior-subordinate feelings). The history of oppression lives in the present, and regardless of status attainment (including the highest offices in the land), prejudice, discrimination and repression feature across distinct population groups. Such social processes are at the intersection of class, race, ethnicity and gender. They materially cut across class divisions and hierarchies even while being framed within overarching class relations and interests (e.g. there are intrinsic links between capitalism, imperialism and the notion of the 'white man's burden'; women at home have long constituted a 'reserve army of labour').

Acknowledging and responding to the culturally lived experience of being a working class or middle strata member of the black community or a working class or middle strata woman is therefore a *necessary condition* for decisive political change. Such engagement is an integral part of the transformation of the ideological componentry through which systemic consent is secured.

The appearance of greater equality of opportunity across class structure carries considerable weight in securing consent. Certain cross-class identities that flow directly from the historically institutionalised expressions of capitalist hegemony, such as *nationalist identity* accompanying the formation of nation states, are also deeply embedded across working and middle layers of the class structure. It is an integral part of the cultural fabric through which class location is lived.

We mentioned above the cultural and legitimated ideological weight of 'private property and capital ownership', whether and to what extent one has this in actuality or as aspiration. To ideologically fracture legitimated consent for private ownership and control of the means of production and accumulation cannot be done independent of embracing the distinction between the private property and capital ownership of subordinate classes from that of the ruling class.

Likewise, as mentioned earlier, a *culture of democracy* exists across subordinate classes in the advanced economies. Not only was democratic franchise won 'in struggle' by large sectors of subordinate classes, democratic culture also became historically embodied with the creation of trade unionism, public literacy and education, freedom of movement and even more dramatically now with the advent of ubiquitous social media access and expression.

Even with the relative decline of unionism across the working class, and the opting out that often occurs in relation to exercising the right to vote, the culture of democracy within the liberal capitalist democracies is a significant demarcation from that which existed historically as a pre-condition to the revolutionary transformations in Russia and China. The Stalinisation of these revolutions was in no small measure due to the absence of this democratic culture amongst subordinate classes (especially the illiterate peasantry) in the revolutionary transformations.

Democratic voting culture is important within the capitalist democracies because it actually sits in contradiction with the authoritarian structure of capitalist ownership and control of the economy and the Capitalist State that is grafted onto that ruling class structure. Voting every three or four years for various levels of government is often portrayed in the revolutionary socialist left as part of the mythologising of capitalist democracy and contrary to the genuine democracy that should be pursued as part of socialist transformation. But there is more to it than that and it would be (and has been) a political mistake to discount its significance.

The actual gains of participatory liberal democracy are, of course, limited in form and content of application, especially in the context of authoritarian capitalist ownership and control structures. Socialisation of the central means of production and appropriation of the social surplus means precisely to democratise these authoritarian capitalist ownership and control structures. Actually contesting the constraints within existing forms of liberal capitalist democracy, plus building upon and extending them, is an important part of the political equation for socialist transformation. Most decisive however is integrating the existing lived democratic culture of subordinate classes as part of the intrinsic ideological narrative of socialised transformation of capitalist ownership structures. To not do that, as for example with the left social democratic 'tax and regulate' GND, is to effectively circumscribe the parameters of reform or accept if not capitulate to the authoritarian structure of capitalism rather than its democratic transformation.

Comparable considerations apply to culturally embedded national identity within the historical contours of nation-state formation. The omnipresence of the capitalist nation-state (economically, institutionally and linguistically), and its integrated cross-class 'relative autonomy' developed over hundreds of years of development, weighs heavily on the nationalist culture and ideological base and expectations of subordinate classes.

Cross-class identities like *nationalism*, superimposed upon the foundation of class structure, have been a critical condition of capitalist hegemony.

Nationalist ideology is used by the State apparatus, various capitalist interests, pro-capitalist political parties and the Ultra-Right to galvanise subordinate class consent around particular self-interested agendas, whether it's a
mobilisation for war, against foreign corporate capital competition or invasion by economic refugees. The success of this discourse is predicated upon a
nationalist culture being historically and materially embedded across subordinate classes, structurally and inter-generationally. Nationalist culture has
become a significant perceptual lens through which 'individual' and 'group'
political interests within class structure get filtered and shapeshifted. This has
a significant material ideological effect on the political terrain that needs to be
negotiated on climate action.

It is important here to distinguish between nationalist culture and
nationalist ideology. *Nationalist culture* has emerged over time and gets
embedded and transferred inter-generationally in everyday language,
customary practices and identifications of self, community and other ('I am
Australian'). As with the other aspects of culture relating to family relations,
or religion, it exists at an often unreflective conscious and sub-conscious level
of acceptance of ideological and historically constructed identity, social
relations and practice. To borrow a formulation from language philosophy,
we think we are tracing around the nature of 'a thing' but we are merely
tracing around the (cultural) frame through which we look at it.

Nationalist identity, in a cultural sense, often exists in its myriad forms as
not much different from a sense of community identification ('we are Australians'). It is embedded and reproduced as much by one's sense of
engagement, dependency, and future well-being as integrally tied to the
national economy, the nation-state and the 'community of citizens' within. It
is reflected and reproduced at such simple and mundane levels as barracking
for one's national sporting team. It also is at the heart of passport politics. In
this sense, it is an identity that is bestowed, by birth or by naturalisation.
Citizenship and entitlement are integral to concepts of nationalist identity.

Everyday language reflects and reproduces cultural practice. Sexism,
racism and exclusionary or privileged nationalist identity live within everyday
language and culture to varying degrees of intensity. In the case of national
culture, self-identity, community and difference to 'other' is a malleable and
contestable space at the same time that it intrinsically embodies resilience and
rigidity in the continuous stream of day-to-day reproduction. This unreflective cultural identity and practice, such as how we filter and express our

relation to raising sons and daughters, our view of (non-English speaking) migrants, refugees, Indigenous peoples, or supporting 'our' national sporting teams coexists with a sense of cultural identity and difference based upon race, ethnicity, gender and/or class.

This inherited and existing mix of national identity and difference is distinct from the *ideological mobilisation* around a specific aspect of national cultural identity. There is a difference between support for a national team and the tribalised nationalism of soccer hooliganism between groups of supporters, particularly its pre and proto-fascist versions. Much of national and nationalist identity, like with racism or sexism or aspirational property acquisition, is deeply embedded at sub-conscious cultural levels as well as at more spontaneous and unexamined consciously expressed levels. It is entrenched in and through everyday language and via common repetitive practices and interactions. Much of this is simply inherited through being integrated into an existing culture from birth or through community social-isation or social learning processes associated with formal immigration.

When reflectively or analytically examined these lived cultural ideations and practices can be revealed and understood for their spectrum of unexamined, unintended, socially backward, ignorant, destructive, oppressive and/or reac-tionary dimensions. Not to throw the baby out with the bathwater, they also can be intermixed with socially progressive elements and aspects (e.g., work-place and community solidarity, volunteering to help others). This can and does happen simultaneously. This feature of national identity as social practice means that this cultural expression does not particularly or necessarily inhabit a left or right political universe for those who inhabit that cultural space. That is the domain of ideology, which mobilises specific forms of cultural identity and action around an articulated set of precepts and agendas.

Differences are not divisions, although they are frequently used in attempts to build walls to separate and divide subordinate classes and social groups. They are also possibilities and opportunities to acknowledge commonality in diversity, to reach across and build a community of shared interests. This search for unity in and among our inherited and entrenched cultural differ-ences applies to our political approach just as much as it applies to everyday cultural life in a class-divided society. Acknowledging the cultural reality of lived differences (in all their complexities, including the confronting expres-sions) is a platform to re-construct the tendencies to division rather than to concede or relent to those tendencies.

Ideological demarcations that accentuate subordinate class insecurities or their relative advantage (compared to other subordinate class sections or social groups) are perpetrated by sections of domestic Capital and/or the Ultra-Right. This is seen in efforts to demonise minorities of colour, refugees, or foreign capital, or 'global government' international agencies, or integrating subordinate class 'shared interests' with the self-interests of national capital. But all of this is a politically contestable national cultural space. It must be countered by exposing, for example, the actual private interests rather than inferred common national interests of domestic capital. It needs to be fractured by asserting democratic demands that prioritise the national majority *community public good* of subordinate classes, within the national economy and in the international context, against the private interests of domestic capital.

This doesn't obviate the political difficulties and messiness in constructing a national counter-narrative, or in adopting a tactical and strategic approach to accepting the democratic constraints and possibilities of a particular question. For example, a thousand refugees at the border present different democratic national political challenges from a million. A Left political principle about having 'open borders' because 'we are internationalists' effectively abandons the national democratic challenge and the political contest to the political Right. Placards and slogans demanding that we fracture majority national consent to off-shore refugee incarceration camps are a poor substitute to actually democratically winning opposition to refugee incarceration in advance.

Fracturing consent requires a politically nuanced approach to the existing cultural base through which class conditions are lived and filtered and the ideological narrative through which it is democratically mobilised and transformed. It's the difference between American gun-owning hunting culture still finding expressed merit in the banning of military-style weapons versus attending anti-pandemic lockdown protests with a rocket launcher slung over your shoulder (yes, for real). It's also the difference between a sense of national self-identity, community and pride (as in responding with donations, recovery assistance and policy responses to national crises such as bushfires disasters) versus support to pursue an abstractly framed national/patriotic interest in a domestic/overseas war on terrorism or attending rallies against Islamic mosques and schools.

The political challenge is also reflected in a 'left nationalist' trade union discourse to resist competitive disadvantage generated by multi-national corporations, and by domestic capitalist importation of 'cheaper migrant labour'. In Australia, a de facto 'White Australia' migration policy was pursued and secured by the trade union movement from the late nineteenth through to the late twentieth century precisely to that end (in the process reflecting, reproducing and further entrenching deeply ingrained racism in popular culture). More generally, such trade union 'left nationalism' is often integrated with political attempts by powerful domestic capitalist interests to mobilise and ally national labour interests with their own against 'foreign' Chinese capital or re-negotiating multi-lateral trade agreements for the sake of 'American workers first'.

The political traps of a nationalist discourse are most often accentuated not because it is nationalist in itself but because it is articulated with the preservation of the existing structure of power. The left nationalist trade union illustrations above reflect that trade unionism itself is centred upon negotiating a greater balance in the distributional outcomes from the existing structures of power, be it the policy parameters of the State or the wages and conditions offered by employers (individually or collectively) or clinging to 'Australian jobs' in climate damaging industries. In so doing, they reflect limited parameters and further quarantine an agenda for transformation of the structure of power itself.

A left nationalist discourse for the democratic transformation of domestic capitalist ownership and control, and the State which oversees its interests, also transforms the political parameters of nationalist discourse and is what gives it a progressive dimension. To draw a parallel, a nationalist discourse for national self-determination against colonisation and/or imperialism is progressive insofar as it is both a democratic expression against an authoritarian external force of occupation as well as the constituent components of the actual structure of ownerships interests carrying out the economic exploitation. In this sense, it constitutes a national-level struggle for democratic control over both State (relative to external governments) and Economy (relative to elite class interests), which in each case is profoundly anti-authoritarian.

This also illustrates deeper dimensions. Imperialism granting, or being forced to concede, some measure of democratic national self-determination (as occurred with post-war de-colonisation) ensured that this happened within limited parameters in many instances. De-colonisation often occurred without altering the fundamental relationship of accumulated plunder of the national economy by dominant imperialist capitalist ownership interests. For instance, it frequently involved continued collusion between external ownership interests and local ruling class/caste interests. This frames how the structural duality of democratic nationalism and economic ownership can be severely circumscribed. The experiences of post-war African countries with de-colonisation and self-determination illustrate how a democratic nationalism without national economic ownership is more than a pyrrhic victory yet also a great deal less than what was on offer.

A democratic nationalisation of the constrained structure of the liberal democratic capitalist state must be part and parcel of the democratic nationalisation of the private ownership structure of the economy that the existing State is built upon. This important theme is discussed further in our concluding chapter.

FRACTURING THE HEGEMONIC CENTRE

In lived experience and by definition, culture is predominately conservative amongst subordinate classes. That's because it is a set of inherited and structurally reproduced identities and practices built upon dependency and insecurity. By 'conservative' we mean more than politically shifted to or embedded in a *right* discourse or disposition. Everyday culture may combine elements of right and left. It depends upon how these are defined and interpreted. It is illustrated historically above by racist elements within trade unionist approaches to perceived competitive threats from migrant labour as well as with the gender-exclusive robust masculinity of traditional manufacturing unionism in Australia.

While conservative, subordinate class culture is also malleable. It is not intrinsically right or left nor immutable over time. It rests upon a foundation of subordination – namely, capitalist relations of exploitation – that renders it particularly sensitive to conditions of precarity and sudden change.

The principal point is that the cultural conditions of class and cross-class identity, through which material class conditions are filtered, are an especially contested political and ideological space in times of crisis. They form the basis through which ideological transformation needs to be shaped in material conditions that demand transformational action. The Left needs to embrace the challenge of the ideological contest or it will be run over by the trans-formational imperative of economic and ecological crises. Reactionary political forces have a structural advantage of existing power and control that enables a 'preservation of shared interests' narrative with respect to insecure subordinate classes.

In the case of climate action, which requires urgent and fundamental transformations of 'old economy' extractive industries, the imperatives to democratically win majority support for this transition requires fracturing the relationship of immediately shared interests between Capital and Labour in the ongoing viability of the enterprise. As with the question of nationalist discourse above, unless climate action solutions provide a secure transformation that includes the interests of labour and dependent communities then the political support for extractive capital interests is strengthened by default. This is doubly ironic in that the interests of extractive industry capital, which are already focussed on labour-shedding capital intensification, are increasingly moving in the direction of capital flight altogether. Both imperil the interests of employed labour and local communities in the immediate term in any case.

Fracturing of the conscious cross-class consent of middle layer 'capitalist friends with benefits' in periods of structural crisis tends to more, not less, extremes of nationalist identity and underpins the success of Ultra-Right narratives in the current period. The institutional relative autonomy of the capitalist nation-state is the only leverage that dependent middle strata have to counteract the negative effects of global capitalist interests, crisis dynamics, and power of dominant interests to impose their accumulation crisis solutions.

Growing middle strata discontent expressed through an Ultra-Right nationalist narrative is integrating with growing layers of working-class discontent against globalisation in the advanced economies. Neo-liberal dismantling of public ownership and the social democratic welfare State over the past 40 years, globalisation of manufacturing, and stagnation of wages has eroded the materially advantaged conditions of workers. Also eroded has been the numerical and political strength of trade unions that delivered these conditions. The contrasting proportionate growth of contract and part-time labour, the gig economy, and the expansion of consumer and housing debt has made conditions of job security and relative prosperity even more precarious. The deepening labour force effects of the accelerating technological revolution in information technology and robotics will exacerbate this discontent in the decades ahead, as will the structural costs of climate change adaptation and mitigation.

Over the past 100 years the prevailing scenario in the advanced economies is that we have had the richest poor people the world has ever seen and its most historically affluent labour force. In comparison, for hundreds of millions in the less developed sectors of global capitalism, life was truly awful and then you died.

Globalisation has dismantled remnant pre-capitalist modes of production and built a cheaper manufacturing labour force to that on offer in the advanced economies. Cheaper mass consumer goods and debt have buffered the structural decline in relative prosperity of labour in the advanced economies and especially in relation to that of global labour.

At present there are some 6 billion people globally that are still aspiring to the conditions of existence of the 1.5 billion fortressed in the advanced sectors of the global economy. This forms part of the unsustainable accelerant of the ecological extinction path. It also forms part of the 'threat dynamic' feeding into the insecurity and structural decline of relative historical advantage within subordinate classes.

Consequently, neither outright neo-liberal governments nor the 'Neo-liberal Lite' social democratic versions are managing to retain ideological and political hegemony over those 'left behind' by the globalisation of the dominant capitalist interests these governments serve. The fracturing of the hegemonic centre of advanced nation-states within the globalised economy continues to deepen. Acceleration of forced migration by Les Misérables outsiders to the advanced economies forms a critical political narrative for the Ultra-Right nationalist agenda to integrate the insecurities of the growing ranks of malcontent subordinate classes.

Indeed, and undoubtedly, ultra-nationalist narratives represent a particular challenge for constructing a transformational counter-narrative and political force for the Left in the advanced economies.

The structural propensity of subordinate classes is to demand the State mediates and buffers the consequence of economic crises and secure their immediate material interests and well-being. It is a conservative political propensity in the first instance, insofar as it constitutes demands *from* the existing system of power. The initial conservative propensity of subordinate classes accounts for the comparative ease of success of Ultra-Right narratives over transformational narratives.

Ultra-Right narratives rest upon the *distributional threats* (real, imagined or fabricated) to the precarious and dependent structural advantage of subordinate classes. Anti-elite rhetoric, framing national and ethnic betrayals, scapegoating and externalising threats all feed into the material insecurity, vulnerabilities, and exclusions of subordinate classes. This is especially the case in economic and political legitimacy crises. But, however much the Ultra-Right rhetoric is about 'burning the house down', it's not about structurally replacing the owners but about replacing the politically bankrupt house-keepers that run it.

Transformational narratives rest upon contesting and *replacing the existing ownership system and relations of power* themselves. It's a steeper political hill to climb. But it is just as steep, and especially politically fraught, for whatever versions of social democracy one cares to embrace. This is particularly so in the face of an Ultra-Right onslaught in conditions of crisis. Social Democracy at best offers a different form of institutional management and more equitable distribution flowing from the existing structure of power. In the context of 'social breakdown' political dynamics it turns into a 'trust me' defence of the core structure and institutions of existing elite power. Social Democracy is political roadkill for the Ultra-Right in periods of systemic crisis.

The current pandemic economic crisis is of such a scope and scale that all the pre-crisis political struggles and options are being re-cast. If the crisis is not on track to match or exceed the Great Depression, then it's projected to go close. It is certainly exceeding the scale of the GFC.

This makes it a particularly critical time for the climate movement and the Left generally. What mass momentum there has been has now stalled, as has the left social democratic electoral push for a Green New Deal (although the extensive and persistent Black Lives Matter demonstrations in the United States highlight that strong underlying discontent is still capable of growing into a systemic transformational force within the immediate period). But some things are clear:

• The current economic crisis will be deep and long lasting on the scale of the Great Depression, especially for the more economically vulnerable and marginal layers of subordinate classes, including middle strata.

- There will be greater fracturing of consent with the existing structure of power and their legitimising expression in the political apparatus of the State. This will fuel the very real risk in growth of Ultra-Right narratives among the increasingly dispossessed 'advantaged layers'.

- Growth in 'legitimated' authoritarian dispositions on the part of liberal democratic nation States in relation to the pandemic, general social order, and who is to pay for the budgetary crisis over time is set to continue if not escalate, along with yet greater fortressing of borders and people movements generally.

- There will be greater political push-back by vested interests to the re-structuring cost of greening the re-start of growth and accumulation, especially when controls have been relaxed during the pandemic period (as per suspension of EPA pollution restrictions in the USA).

- The risk has escalated of the green movement, left social democracy and the anti-capitalist Left generally being as 'left behind' as our subordinate class base. Unless these current emergency conditions become the basis for coalescing around a common program and connecting with the class forces required to create a transformational movement, then a more immediate barbarous course looms.

The good news is that amid this globalised crisis there is a political opportunity to review, reset and rebuild a movement that has been fragmented and floundering. The contradictions of capitalism also embody the seeds of its destruction. This provides an opening that must be seized by the broad Left.

5

REBELLING FOR A GREEN CAPITALISM
IS A DEAD END

In this chapter we address one vital question: can the capitalism that brought us to this climate endgame be turned green? It's the central question of our historical moment.

For our movement, the unfolding climate catastrophe has provoked much angst and imaginings for alternative pathways to address the inexorably growing emissions. There are a multitude of political, economic and technological reforms being proposed to adapt, mitigate, cap or reverse the consequences of climate change. Many of these are responding to the climate science, which identifies key trends and problems, but which often leaves open the political question of how implementation to contain heating is to be achieved.

On the other side, there are also a not inconsiderable range of such measures being implemented and foreshadowed by those in charge. Some on this side have not ignored the science either. The predicament of the extinction curve being propelled by the capitalist growth imperative is now also fracturing the wall of silence and resistance of our ruling class. Hope and promissory notes increasingly emanate from the corridors of power along the lines that reforms of ecological and economic malaise are possible – that sustainable endless growth can continue, with prosperity for all, albeit powered by technological optimism and more renewable fuel.

At one level this is a welcome change from the fabricated climate change denials, minimisations and contrarianism still spewing from neo-liberal think-tanks and Fox media outlets. These are precisely intended to favour and protect the interests of specific corporate vandals and the system as a whole.

But these obfuscations contesting the climate science are being overwhelmed by the everyday realities and 'truths' of global heating and its consequences. The power game is now rapidly shifting to one of governments integrating the discourse of urgent climate change response into the existing political and economic global order. This is happening even while subsidies to the fossil fuel industries continue, and restrictions on their polluting activities are being diminished due to the coronavirus pandemic. But the general movement is nonetheless clear.

This dangling carrot of hope for a green capitalism is ironically being mimicked in the still rather polite protest marches and policy petitions to the very doors of the power elite – the corporate boardrooms of transnational corporations and their national and international state political institutions and agencies. There are a multitude of regulatory reform agendas for a more ecologically sustainable capitalism, or a more socially equal capitalism, a more productive waste-managing circular capitalist economy, or even a post-growth capitalist prosperity. Activists are even being offered training by other activists in how to speak to board members about these exciting new options.

Most of such aspirational, eco-capitalist scenarios are built upon foundations of democracy, more equitably shared resources, a green industrial revolution, and a more balanced and harmonious relationship with the natural world. But can we truly embrace the prospect of a future eco-capitalist mode of production, one that would simultaneously and altruistically re-configure growing inequality and disproportionate collective ecological impact all the while leaving the core structure of private ownership, accumulation, control, and political and social power intact?

We consider such forms of hope for eco-capitalist reform are badly misplaced at best. At worst, 'imagining' or 'finding' or 'clinging' to such hope being realised becomes the rope that will hang us all.

A CONVERGENT MOMENT

The global financial crisis (GFC) laid bare the fragility of the edifice of capitalist growth, the fundamental and irreconcilable class interests involved, the stark political choices required, and whose governments 'our' governments essentially are.

This crisis response is being replicated by governments globally in the current 'economic hibernation', with Keynesian-style pandemic budget packages such as here in Australia. As before, corporate enterprises have been quarantined from collapse with public monies. The capital-labour relation is being preserved with government-funded wage guarantees being channelled through employers, so that growth and private accumulation can re-start sooner or later on the same terms. For middle strata businesses and landlords, the government assistance picture points to a much more mixed and murky future. Many will fail permanently or have even greater indebtedness to their bankers and/or large landlords.

For housing renters, casual and overseas workers (one million in Australia), and mortgage holders, the losses and debts remain fully privatised and over-whelming in the immediate term. What little emergency 'hibernation' measures are in place for subordinate classes overall barely rise to the level of basic subsistence. To take one example, the Australian government assistance response for workers is more comprehensive than most advanced economies. Yet within only two months of economic lockdown measures, some 1.5 million workers have had to apply for early access to their mandatory private retire-ment savings totalling over AUS$11 billion. This level of desperation will be magnified by a still unknown but certainly double-digit unemployment across the global economy as a whole.

Moreover, the State budgetary debt consequences of socialising big-business losses will certainly impact on subordinate classes over the longer term. They will be the primary bearers of the tax burden and reduction of State social services (including no doubt the national health services that have been the saving grace for many in those countries that have them). Of course, the current crisis is still unfolding, and it remains an uncertain and contest-able policy space even within the ruling class itself. The current Keynesian response has wide agreement but whether 'business as usual' returns in full neo-liberal austerity mode or morphs into a MMT 'magic money tree' direction is unclear.

In Australia, capitalist economic think-tanks are already laying the ideo-logical parameters and foundations for what the post-pandemic State budget policy landscape will need to look like: a redesign of the tax structure to incorporate higher consumption charges and reduction of exemptions for retirement savings and private housing. No mention of course of higher rates of corporate tax or of nationalisations in the national emergency. Rather, if anything, larger tax cuts for big business are back on the agenda. In the case of airlines, the government is struggling to avoid de facto nationalisation, such is the scale of socialised losses required. Greater interest-free lines of credit and tax breaks are already being foreshadowed to 'stimulate' private capital investment. There may be a MMT for corporate and State debt, but it doesn't extend to the broader non-corporate regions of the economy and the private debt burdens of subordinate classes.

The re-setting of the capitalist economy will also impact on the immediate, limited measures foreshadowed in the transition toward a greener capitalism. The EU's *European Green Deal* transition is already openly being questioned by finance ministers and arguments continue regarding income, expenditures and investments, not to mention who ultimately pays the costs (workers' taxes) versus who most benefits (the private sector). Also in question is the unity of the EU bloc as a whole over the internal nationalisation of the Covid-19 emergency response and accompanying national partition of budgetary consequences and obligations. If taking care of one's own is a refrain that prevails, it is near certain to kill off collaborative green action by governments across Europe. It will certainly feed an Ultra-Right narrative.

Reactionary governments, like in America, are grasping the opportunity of the Covid-19 economic crisis for an accelerated wind back of existing ecological regulations that inhibit business recovery and expansion. Support for the fossil-fuel industry is being ramped up to help deal with the oil glut arising from the pandemic collapse in demand. Meanwhile in Canada, pipeline construction has accelerated in the light of the barring of public gatherings thereby stifling and stopping protest movements in their tracks.

On the other hand, the current crisis has also revealed some positive potentials and opportunities for ecological action. For instance, it has concretely demonstrated what interrupting 'business as usual' can actually mean for the climate emergency as witnessed by the dramatic improvement in air quality globally. Also demonstrated is that a massive global economic re-organisation is possible, that monies are available if the emergency and the public political will is significant enough, and that reversible ecological impacts of a flattened growth curve are achievable and can be immediate.

The problem is that the re-set opportunity of the current crisis, and the potential lessons it provides, requires a movement of political clarity to galvanise democratic support around this opportunity. As with the GFC aftermath, this clarity of political purpose for our movement continues to elude us. But for those in charge, the purpose and intent are clear. Business as usual will restart as soon as possible.

As noted in Chapter 3, endless commodity production for private capital expansion and profit accumulation lies at the heart of the unfolding ecological catastrophe. Growth is a systemic imperative, not a policy choice. Competing and irresolvable contradictions embedded in the core dynamic of capitalist accumulation will continue to lead to inevitable and recurring systemic crises. The scope and scale of the current crisis is both a convergent generational political moment and a projection of futures to come.

The incapacity of capitalist growth to deliver on the promise of sustainable global prosperity for those of us subjected to its dictates has never been more transparent. Advocating that we can just moderate the terrible social and ecological costs, or that the core structure of capitalist social power and self-interest can be simply negotiated or circumvented, is a vain hope born of a Faustian bargain. Dealing with the devil only ever has one result and it is not a good one for us. Forging a path out of this political cul-de-sac is the central historical challenge for our movement and subordinated classes globally.

TIME'S UP

After 270 years of capitalist fuelled expansion of production and consumption, climate catastrophe is now already inevitable. Scientific analyses of climate change point to two major trends. First, the planet is heating up faster with each passing day, and greenhouse gas emissions continue to rise despite scientific warnings and political commitments. Second, each refusal to act decisively and immediately exacerbates global heating. Delayed cuts in greenhouse gas emissions means ever deeper cuts are needed as we run out of time. The period between Kyoto and Copenhagen was the last (barely credible) political opportunity to create the time buffers required for more sustainable global economic transitions. Even then, it has only really applied to the energy system fuelling global capitalist expansion, the fossil fuel industries. Now it is only the *scale of the catastrophe* that is at issue. The very conditions of capitalist power and interests that have resisted a sustainable transition over the past 40 years have now made the political prospect of a sustainable eco-capitalist transition impossible in the remaining time frames available.

At the current rate we will be at 450 ppm of atmospheric CO_2 in about 15 years. This means that limiting global heating to 2 degrees above 1750 levels is unachievable. As Bill McKibben (350.org) says, 'do the math'. That broadly represents a geological tipping point for accelerated feedback loops – elevated release of other greenhouse gases (especially methane), carbon sink overload on land and in oceans, and reduced albedo (reflected light) from irreversible melting ice sheets. Once the self-expanding greenhouse dynamic is unleashed then we will be hurtling toward a new, beyond human control, post-Anthropocene path to a new Venutian Epoch along the lines of our sister planet. It is now structurally impossible to sufficiently re-engineer the global capitalist economy within this timeframe: not in energy, food, water, priority needs-based commodity production, infrastructure adaptation or climate mitigation. Implementation of the emissions objectives of the Paris Accords will not achieve this. In any case, Paris is already heading the way of Kyoto and Copenhagen, and Glasgow certainly holds little promise coming on the back of the global pandemic.

Presently, the global atmospheric CO_2 is at 418 ppm and rising. The current Covid-19 economic shutdown may temporarily flatten that rise this year (2020). But it won't alter the fundamental trajectory. Currently, we are nearly 50% above 1750 industrial revolution levels of 280 ppm. That took 270 years. At current pace that initial industrialisation level will be doubled to 560 ppm in only another 60–80 years. Yet, 450 ppm alone will be an accelerant level for tipping points that no amount of geo-engineering will contain, never mind reverse. If we add all greenhouse gases together, and expressed it as a CO_2 equivalent, then the current level is over 500ce ppm. We are already out of time.

The window to realistically prevent catastrophic levels of global heating was closed some 10 years ago when the GFC failed to translate into a transformable, post-capitalist movement. Yet even now our movement for climate action remains politically shackled to the very structural foundations that have given us this crisis. Neither the opportunities nor the lessons from 2008–2009 have been grasped.

Consider this. There are few plausible green capitalist scenarios for effective levels of containment, mitigation or adaptation. Not for vast layers of the global poor and working-class layers. It bears repeating here that there are some 6 billion people who are aspiring to the conditions of existence of the advanced economies. Of these, some 2 to 3 billion are living in conditions at or below the basic levels of the more marginal layers of our class-divided societies. If, as capitalist growth imperatives dictate, expansion of commodity production fills the market gaps of the material conditions of life of these 3 billion then we would need four planets, maybe more. Globally, we are already well onto our 2nd planet in terms of resource exploitation impacts.

The strategies and recommendations on offer are fundamentally coded to assuage concerns amongst those who feel buffered by living in the advanced sectors of global capitalism or are employed in the more affluent middle layers. Terms such as 'mitigation' and 'adaptation' have become part of the diversionary lexicon as the corridors of capitalist power have progressively shifted from outright denial and 'climate scepticism', to a grudging 'accepting of the science', embracing the 'science of geo-engineering', and promoting the entrepreneurial opportunities of 'greening global capitalism' itself. For example, geo-engineering, such as with solar radiation management, is largely untested and encompasses considerable risk in side-effects and long-term unintended consequences amidst the huge technical challenges. Efforts directed at carbon dioxide capture and storage, not just to retain fossil-based energy but to reduce existing CO_2 levels, are puny in comparison to the causal forces of emissions before us. Yet, they still seem to resonate as potential saviours of the moment, both in the public domain and within our movement.

Despite widespread recognition across our movement that time is up, we still seem trapped in the same political logic of capitalist reform. Rather than a unified political hammer that locates the climate emergency as a specifically systemic capitalist creation, 'we humans' now have our own Anthropocene Epoch. Not only are 'we' all culpable but we are 'all in this together' (as the framing of pandemic response also keeps reminding us). This extends to future planning as well, as we are exhorted to think beyond so-called 'political tribalism' and embrace solutions that 'work for everyone'.

After all, we have our geo-engineers to fix the 'oops' of 50 years of wilful denial, eco-modernists to help chart us out of our destructive 'old fossil fuel economy' ways, environmental criminologists to help re-purpose the regulatory and oppressive apparatus of the capitalist state, carbon market proposals to make green-transition capitalism 'work for us', and Green New Deals to channel dissent into restored, revised, pre neo-liberal, social democratic reform agendas.

We should add that we ourselves have not been immune to these tendencies. But this is precisely what needs changing at precisely this moment. The time for tinkering is past. Also past is holding on to illusions that fundamentally changing the system from within is somehow possible. No matter how loud our voices or how strong the argument or how rational the policy, those in power have interests that diverge from the rest of us. They will not just give them up.

Capitalism has its interdependent and contradictory fragilities, but it is nothing if not historically and structurally robust. Leveraging off crises to find sustainable terms of consent from the governed is the political forte of the capitalist State. Emergency powers and strong governments give the appearance of united interests in the same moment that they prop up the business end of town and degrade workers' rights and environmental protections. Retaining support for continuing private accumulation is the art and part of the very mechanism of dominant class power. Accordingly, striving for a greener capitalism as a solution to the climate crisis that capitalism itself has created is but a fig leaf for the extinction path. It does not address core issues. Rather, it reinforces capitalist hegemony which is premised upon knitting sufficient ideological consent across contradictory and divergent interests within an overarching framework of minority class control.

ENDGAME

Unsustainable, unfettered capitalist growth has to come to an end. Even the most dinosaur elements within our ruling class can see the ecological meteor coming. But like dinosaurs they don't think it really applies to them. The thing is…they are fundamentally right.

Subordinate classes may get confused and conflicted about where their core interests lie, but that doesn't apply to the ruling class. There are the super-rich that don't care, and know they are buffered by whatever calamity may unfold in their lifetimes. Then there are the super-rich that are philanthropically inclined because it barely dents the money capital at their disposal.

Some, like Gates, create foundations to altruistically solve a particular social problem. Buffet has even called for the rich to be taxed more than the 3% he feels obligated to pay. Bloomberg, with barely a blink or hint of irony, offered to buy the Democratic presidential nomination and subsequent election as a public service to clown-proof Washington. The initial down payment was $60 million. His personal Super Bowl ad was $11 million. Financier George Soros is the bête noire of the Ultra-Right for his liberal causes.

The Koch brothers (one just died in 2019) have also been philanthropically inclined. They, along with Murdoch, have been the most conspicuous political funders of the Ultra-Right think tanks, media and super-PACs that underpin the climate denial industry and the Republican political machine. They aren't really climate science dinosaurs of course. The Koch family and their less conspicuous contrarian funders know full well the impact of science-based green transition measures on the potentially stranded vested interests they hold.

The fossilised Republican capitalist wing and their minions may sound and act like raptors, but their mammalian rat cunning comes with cutting edge political tech. Data mining of social media outlets has allowed micro-targeting of particular constituencies with social/cultural issues polling, political ads and misinformation campaigns. Tactical electoral voter turnout or suppression (not to mention Republican State gerrymandering) translated into the election of Trump with 3 million fewer votes than Clinton. This strategic and tactical political nous was also evident in Brexit and underpins the continuing rise of the Ultra-Right on both sides of the Atlantic.

Radical conservatism has been entrenched in the USA for decades. They have the clarity of agenda, the money, the organisation and the levers of political power to prosecute the interests that lie behind them. But it's not all of one type. It has always been a coalition of overlapping and competing conservatisms. Correlations with the direct interests of different wings of the American ruling class is not empirically straight-forward. Moreover, it is built upon and shifts with the cross-class consent requirements of subordinate middle strata layers in particular.

The radical conservatism of the Reagan/Thatcher neo-liberal revolution through the 1980s fused the globalising interests of American and British/European capital (with 'free trade' reflecting their competitive strategic global advantage) with the de-regulated small State, low-tax, anti-union domestic agenda that particularly appealed to the propertied interests of growing middle strata. It had a dramatic conservatising effect upon the traditional reform alternative of social democracy across the advanced economies.

This neo-liberal lite conservatism was incorporated into the political agendas of Australian, New Zealand and UK Labour parties, the left US Democrats and the European Social Democratic parties, as reflected for example in the 'reform' governments of Hawke-Keating, Blair, Clinton and Obama. It was this that paved the way for Trump, Brexit, and the rise of the

Ultra-Right across the advanced economies. It left behind the immediate material interests of its traditional working-class base for the priority interests of managing a globalising capitalist economy on behalf of Capital. The political pitch of social democracy was re-adjusted to focus on the growing affluent and aspirational middle strata layers of the working class. It has created deep political fractures within and between the working class and social democracy across the advanced economies. It was on full display in Brexit and in the resulting fractures between Leavers and Remainers that politically paralysed the British Labour Party.

The current second iteration of radical conservatism is built upon the crumbled foundations of neo-liberalism and globalisation which culminated in the GFC and its aftermath. Tea Party Republicans have largely recovered from multiple, U-turn induced, whiplash events of the 'debt and deficit' Wall Street bailout Obama inherited from neo-liberal Bush and his Goldman Sachs advisors. If not rhetorical embrace, there is certainly a deafening silence on the continued money printing and elevated national debt under Trump. In any case, political whiplash for Republicans and Fox News is now a condition of existence in the 24 hour Trumpian Twitterverse. This darker underbelly of crisis-generated radical conservatism has deep links into the disaffected layers of middle strata and the working class. The victims of the diminished advantages of capitalist globalisation are directly integrated into the strategic competitive pressures of sections of capital in their response to that same globalisation. Cross-class ideological mobilisation – anti-China and against other competitive capitalist blocs such as the EU, anti-multilateral free trade agreements that have moderated previous economic advantage, anti-globalised institutions with regulatory constraints and oversights balancing competing global capitalist interests – are all built upon a more forceful and belligerent use of nationalism and national power to restore diminishing competitive advantage.

The Ultra-Right transformation agenda is our core political challenge. It is built upon a cross-class crisis dynamic that is directly opposed to the centrist, liberal capitalist, pro-globalisation elites. Our core movement is currently trying to ally with this globalised regulatory elite in support of a green capitalist transition agenda. One consequence is that we are doing this without any significant support from the subordinate classes which are being materially disadvantaged if not disenfranchised by this globalisation.

A critical democratic weight is required to give our movement the political clout to challenge the business as usual agenda of the globalisation wing of ruling interests. These subordinate class democratic battalions are still in play. They have immediate material crises and strategic class interests that can be organised around a progressive transformational agenda. Think here of farmers mobilising to protect local environments and scarce water resources from predatory miners and agribusiness, and middle strata businesses being crushed by banks and large landlords in the current economic crisis. But they are also being coalesced around a 'transformational' Right political trajectory.

Our potential supporters largely distrust if not hate the pro-globalisation elites and their centrist political wings. This includes neo-liberal lite social democratic versions looking after national and global big-business interests that are democratically in conflict with our potential supporters' immediate interests. Distrust of detached and distant political elites (in Washington, London, Brussels, or Canberra) matches subordinate class suspicion of democratically unaccountable economic elites (especially foreign owners). It's a contradictory and strategic dilemma that is politically paralysing the climate action movement and the broad left as a whole.

What unites the centrist pro-globalisation elites and other disparate wings of the ruling class, whether they are actually climate science dinosaurs or not, is that their ownership and control of the central levers of the economy and the State apparatus has to continue. The partial competitive fracturing of the ruling class around vested individual and corporate interests overall still remains bonded by this shared class glue. The wall of silence and resistance to acknowledge and respond to climate change may be breaking down but not the structural fabric of private ownership and class rule. There may be arguments about the solution, but whatever the proposed solution it is ultimately a capitalist one.

The acknowledgement that unsustainable capitalist growth has to come to an end is not because of any moral qualms on the part of the globalist wing but because the reality of climate change and growing ecological catastrophe is directly impacting on a greater range of their ownership interests. Ruling class fractures over climate change acceptance or denial reflect the divergent interests flowing from their ownership and control of portfolio assets.

For instance, energy companies are choosing to not reinvest in modernising their ailing coal-fired power stations because of the risk of having a stranded asset. Banks are becoming reluctant to provide loan capital to fossil-based enterprises for the same reason. Even the Saudi Royal family is diversifying beyond their lake of oil.

Insurance companies fear being buckled under by the mounting weight of claims from climate disaster events. This is reflected, for example, by local municipal councils being compelled to change their planning rules in regards seaside and tree-side residential construction in the light of ocean level rises and bushfire threats. Climate litigation includes action around corporate reporting and forecasting that says it must include consideration of the impacts of climate change, thereby entrenching new obligations to shareholders on the part of company directors. Nation-state governments and their internationalised apparatus (like the World Bank and the IMF) fear the rising budgetary toll of re-construction and the impact of local economic paralysis arising from cyclones, droughts, floods and fire events across wider sectors of the national and global economy.

On the opportunity side of the ledger, car companies are rapidly converting to electric and low-emissions vehicles to fill market demand in the smog-choked mega-cities of China, South-East Asia and the Sub-Continent. France and most of Europe will soon be virtually petrol and especially diesel-free because of legislated restrictions on the old internal-combustion engine. The low emissions energy sector is booming, with solar and wind farms, and battery storage grids. 'Transition fuels' like gas and nuclear are actively being canvassed and supported by former environmental action critics and now include even environmental activists. The revolution in mass food production through genetic modification and the laboratory production of 'meat-like' products is well underway.

It would be hubris on the part of our movement to think that a few demos and street theatrics – even the 6 million climate strikers in 2019 – have shifted the political dynamic forward on a green capitalism that is *against* the will of the ruling class. We are mostly just being turned into foot soldiers for the pro-globalisation elites, their regulatory national and international state apparatuses, and green entrepreneurial sectors in their competitive push to gain a commercial edge. This is evidenced by demonstrations that push for global accords, more privately owned solar and wind farms, and emissions trading schemes, all against the wishes of neo-liberal and nationalist wings of the ruling class, but which do not fundamentally change basic social relations of ownership and control.

Equally problematic, and strategically puzzling, is that the political structure of our 'green capitalist' demands is in fact beyond the capacity of the structure of capitalist power and interests to give. To be clear, this means that we are asking those in power to benevolently exercise power against their direct interests. This is akin to asking the corporate director to ignore the economic bottom line for the sake of the greater good. Their masters, however, are the majority shareholders. Regardless of private sentiments within the corporate sector, it is not going to happen without commercial gain or with commercial cost.

At best we get window-dressing responses. Even with commercial and budgetary manageability over time on matters like energy (as outlined for example in the Stern report in the UK and more recently the European Green Deal) neither the climate emergency timetable, nor the democratisation of social needs production and distribution, nor the wider catastrophes unfolding beyond our advanced economy borders, are addressed. Fortress mentality is reinforced through nationalist exclusion (we protect our own). Greenwashing ensures that we feel good when company reports are provided electronically or we forfeit the Hotel's daily clean towel (everyone is doing their bit for the environment) and together we celebrate corporate-sponsored Earth Days.

Sustainable capitalism has been and continues to be about sustaining capitalism as a priority. If our demands for green economic transformation are unattainable within the timeframe required, and don't require transformation of the structure of power and interests that got us here, then it becomes a pro-capitalist climate endgame by default. It happens on their terms, not ours, and in accordance with their transition timetable not ours or that of the ecological emergency or of the generations following.

The best example of the green capitalist endgame is to be found in Chinese State Capitalism – a government and ruling class that by any measure is impervious to the collective political weight of our climate movement in the advanced economies and globally. Here, several centuries of capitalist industrialisation and the attendant ecological impacts have been compressed into a few decades. They are the largest single national economy expanding fossil-fuelled energy systems. They are also the largest single investor and developer of the green transition economy in energy, building and transportation. They are doing so because these apparently contradictory trends are politically fused around the core structure of ruling interests. Fossil-fuelled development to expand wealth and power through globally competitive commodity production dovetails with the requirement to structurally transition away from the ecological devastation that has come with this development. The latter translates into leading the technological innovation and implementation this green transition requirement demands nationally and globally. The competitive edge to own and control the green transition space for commercial gain and strategic advantage is thus integrated through the direct fusion of the Chinese State with the interests of Chinese capital.

This highlights that capitalism is not a monolith, has the capacity to incorporate and/or reconcile contradictory tendencies, and is resilient enough to change its direction as circumstances warrant (from national to global back to national but global). The broader analytical point is that the empirical form particular capitalist societies take cannot be assumed to act or behave in preset ways or in accordance with abstracted formulations. Each must be assessed concretely, albeit in a theoretically informed way.

To some extent then, the transition to a greener capitalism is already well underway. It is primarily driven by the greater governmental and pro-globalisation elite acknowledgement of the economic challenges and opportunities presented by climate change and ecological collapse. But it's not in the form or timetable that the climate emergency requires. Moreover, the interests that are running this transition are not based upon greater democratic and equitable sharing of economic growth and sustainable ecological impact or the needs of the global majority. They are fundamentally framed to reflect the negotiated advantages run by and for the interests of sections of the ruling class driving the shape of this transition. This fundamentally self-serving and hesitant coalition of green transition interests is nonetheless still under serious challenge from threatened competing interests within the ruling class itself.

Whether the transition pace to a green capitalism continues or whether it gets stalled politically by radical conservatism and/or the economic crisis continues to be highly relevant of course. As a movement we certainly need to strategically demand and *critically* support green technology transition measures that flatten the extinction curve. But this comes with a deadly serious political caveat. As with the transition into the industrial revolution, the technological developments are intimately bound up with the private interests of the mode of production within which it is contained. A greener capitalism led by and for the interests of the ruling class will still mean that, sooner or just a bit later, we are ecologically and politically dead in the end.

To alter that requires changing the ownership, control and policy agenda governing the central levers of the economy. While eco-capitalism will still give us capitalised ecology for profit, eco-socialism will give us a socialised ecology for social need. The 6 million who demonstrated globally in the last climate marches will not be sufficient. The numbers would probably not even be sufficient if they were concentrated in a single major advanced economy, although we would welcome the opportunity to see what we could do politically with 6 million demonstrators in Australia. The reason demonstrating is not enough is that it remains fundamentally framed as a set of capitalist market-based demands being directed at our current ruling class to deliver something back to us. The power and control remain in their hands, as does the self-interested form it takes.

The limited level to which our movement has coalesced around a united transformative national/global political agenda is stark. This limitation goes beyond the smorgasbord of demands across our ecological movement and the disparate voices they reflect. There are plenty of proposals and demands being put forward by climate scientists, other environment professionals within middle strata, green entrepreneurs and political activists. However, to win over the democratic majority required to transform the trajectory of the ecological emergency requires wider integration of the social forces necessary to challenge and replace national and globalised capitalist power.

The current scale of the Covid-19 economic emergency has already altered the momentum for even limited ecological action. It should be self-evident, if there is any lingering doubt, that the class dynamics and consequences of capitalist crises profoundly intersect with the challenges for fundamental ecological action. Witness how quickly climate change exited the stage in the immediate aftermath of the GFC. The same is happening now with the pandemic threatening the global economy. Economic recovery (read capitalist reassertion) is the central focus in the immediate term, both for Capital and the decimated, dependent material conditions of subordinate classes.

The moment for integrating the climate action movement with the pressing economic and political interests of subordinate classes is now. Time's up. The climate and ecological emergency is upon us just as much as the coronavirus pandemic. Incremental reform, re-focussing on the next election cycle, faith in the integrity and resilience of our liberal democratic institutions, trusting that capitalist investment in transition technologies will save the day, hoping for 'our better angels' to prevail against the Ultra-Right or for altruism and philanthropy to sweep across the ruling class...that time is up too.

A CONSERVATISED RADICALISM

The rise of cross-class radical conservatism and its embrace of the Ultra-Right highlight the extent to which radical progressiveness has been conservatised in the advanced economies. There are a number of interconnected factors that have subdued the transformational agenda across the broad Left from the back half of the twentieth century to the present:

- The relative affluence of subordinate classes in the advanced economies;

- Social Democracy's capitulation to and integration with neo-liberal economic globalisation;

- The demise of radical trade unions and decline of trade unionism generally as a proportion of the working class;

- The continued collapse of the traditional revolutionary socialist left that accompanied Stalinisation (the authoritarian bureaucratic Party control of the State and Economic apparatus), and its eventual demise in the USSR, Eastern Europe and China; and

- Single issue progressive reform movements disconnected from a systemic transformational dynamic for alternative government (that is, leveraged policy alternatives and reforms are the key planks of activism rather than challenging the basis of political and economic power).

These intersecting strands conservatising the broad Left in recent decades need to be reconciled if political momentum for climate action is to be widened and accelerated. There has always been fragmentation around the ways and means of transforming capitalist economy and power across the Left. Most of this fragmentation has a genuine basis, some not. Unity has to be forged. It is facile to demand otherwise. Nonetheless, this fragmentation is not the central issue, however convenient it is to justify a demobilised retreat at an organisational or individual political level.

Rather, the rise of a more conservative, minimalist and limited radicalism reflects first and foremost that the Left's transformational agenda, in whatever form, has been disconnected from challenging the core structure of capitalist economy and power itself. Ironically, the Ultra-Right provides a mirror of radicalisation for what the Left has itself become resigned to accept as being beyond contemplation. At the precise historical moment that requires political boldness to address the scale of the ecological and economic emergency, our broad movement is consumed by timidity and political minimalism.

The inevitability of climate and ecological catastrophe reflects the major shortcomings if not outright political failures of the overall Green movement in the past 40 years. It both represents and reflects a wider crisis of the twentieth century capitalist reform project of Social Democracy in the advanced economies, of which the green movement has largely been a constituent part.

By 'social democracy' we refer here to largely organised labour movement based, centrist, political parties (Australian Labor Party, British Labour Party) that wish to offset the imbalances of the 'capitalist market' with a baseline of social provision and welfare netting (such as national health care, affordable education, social housing and so on). It is the 'caring' face of capitalism and generally emerged out of the union and socialist movements of the nineteenth and twentieth centuries. It is essentially a form of social liberalism that acknowledges the government's role in enhancing life opportunities and providing greater equity of treatment regardless of social background, without threatening the basic tenants of capitalism. They started as organised labour political parties with a social compact socialisation objective to compete with the socialist and communist movements, but this receded over the course of the first half of the twentieth century as the latter declined.

The US Democratic Party is not a social democratic party but has features of it (similar to the Liberal Party in Canada). It's not based upon organised labour but does orient to and seeks to get majority support from the union movement and socially marginalised communities. Adopting policies such as the Green New Deal package would have a transformative impact in the direction of a social democratic party. This explains both the attractions of the GND amongst wide layers of the politically activist base and the resistance to the GND within the pro-capitalist Democratic Party machine and elected office holders.

The early priority emphasis of Social Democracy to improve the material conditions and security of the working class is basically over. Buffering the worst excesses of capitalism and achieving greater balance in the re-distribution of capitalist wealth and power, has progressively been displaced by the embrace of neo-liberalism and the priority of managing the National Economy on behalf of Capital.

Maintaining the cross-class hegemony of capitalist rule is the shared core centrist mission with the more conventional pro-capitalist governing coalitions. The 40-year social democratic capitulation to neo-liberalism should surely have removed any lingering doubts. Social Democracy, in political party form, is a form of capitalist governance. It is not an agenda for post-capitalist transition. Even in its most radical early days of collective democratic organisation and forms of public ownership, the limits have always been within the negotiated terms of a privatised capitalist market. It reflected the political parameters and agendas of the trade unions upon which they were based and the contest against the communist movement competing in the same political space. Social Democracy is about wider social distribution of privatised accumulation and political power – i.e., greater balancing of the ratio between wages and profits in the economy and greater equality of opportunity for minority groups to be represented across the unequal class structure.

That is why, in most historical cases, social democratic governments emerge to resolve periods of accumulation and political crisis for Capital. The Great Depression and the Keynesian solution to spend big in order to expand employment is a case in point. It has never been a template to challenge or replace the structure of capitalist social and economic power. The socialisation objective (that is, collective ownership and control of public goods and services), fashioned in competition with the communist movement, has largely been removed in most versions in the advanced economies. For example, there is rarely if any mention of nationalisation in contemporary social democratic party manifestos.

Since the advent of neo-liberalism, Social Democratic governments have largely become a fully integrated part of centrist capitalist government. They have presided over the crushing of radical unionism, privatisation of remnant vestiges of nationalisation in key sectors of the economy (banks, airlines, postal services, telecommunications, energy, water and government services – including health, education, research and roads), and management of a globalised national economy.

For the Greens across the advanced economies, this demise of social democracy as a progressive political force is both a strategic dilemma and a political mirror. As political parties that morphed out of a 'single issue' movement, the Greens largely flourished initially because of the political reform space abandoned by social democracy. The drift of progressive activists away from neo-liberal social democracy to the Greens came without the working-class support base and traditional orientation of social democracy as an alternative party for progressive government. No matter how diverse the collection of progressive issues and constituencies that are assembled under one roof, not having an orientation and agenda for alternative transformative government becomes a conservatising weight to radical change by default.

The same effect is evident in the remnant versions of the revolutionary socialist left. Their entryist strategies to transform Social Democratic Parties from within have largely been abandoned via activist expulsion or revulsion at the neo-liberal capitulation. Some activist layers actually went to the Greens. Some just retreated into single issue environmental activism and social movements like anti-racism, refugee advocacy and LGBTQI rights. For the organisational rumps of the revolutionary socialist left, the orientation to the single-issue movement is not much different, apart from the additional focus of trade union activism in competition with Social Democracy. The flickering hope remains that their time will come if they can just cling to their siloed revolutionary inheritance and replace their membership losses in the interim.

The Greens remain trapped in a 'protest party' state. Green political parties across the advanced economies have struggled to attract more than 10–15% of national support in polling and electoral votes. When access to governmental power has occurred, it has been the result of coalition with larger mainstream social democratic parties. Green issues have been grafted on to wider social democratic agendas. Some progress on ecological issues across the advanced economies can certainly be pointed to. By and large though, green issues are entirely framed within the eco-capitalist space that remains comfortable for social democracy. For the Social Democratic parties themselves, Greens are mostly regarded as electoral competitors for overlapping, single issue reform constituencies, especially among middle strata in urban centres, to be outflanked and neutralised. For the Greens, the disposition is mutual and reciprocated. They don't make a comprehensive pitch as an alternative progressive government to social democracy for subordinate classes as a priority focus.

The conservatising weight of working-class affluence on social democracy's entrenched pro-capitalist governing agenda is now under challenge by the current crisis, as it was during the GFC and its aftermath. There are unique dynamics across various parts of the advanced economies but, without a re-calibration into a party or political force for progressive transformative *government*, neither the Greens, nor the socialist movement, nor our climate movement generally will win the democratic mandate from subordinate classes. It will remain trapped in 'protest petition' and a 'leveraged negotiation' mode within the existing structure of power.

The precarious position of subordinate classes, the more disadvantaged working-class layers in particular, requires an economic security assurance in any progressive transformation political agenda. In crisis periods this material anxiety is accentuated, which is why it's so much easier to politically prey on insecurity rather than 'just trust me' guarantees. The radical conservatism of the Ultra-Right has a de facto head start in this GFC-Covid-19 age of capitalist crises, and this is reflected in Trump, Brexit and the political trajectory across Europe. A minimalist, conservative, reform-focussed or siloed radicalism is politically inadequate for the times we are in.

GREEN NEW DEAL

The Green New Deal (GND) of the left social democratic wing of the American Democratic Party is the most evolved progressive attempt to alter the political contours of the present crisis convergence of economy and ecology. It has its counterparts in the British Labour Party's GND and in the European Green Deal blueprint from the centre of European capitalism. The GND is more 'transformational' in the American context because they are also fighting for social reforms that most of us in Europe, Canada and Australia take for granted.

Social Democracy's earliest emphasis on greater equity won support from the sheer weight of social marginalisation (women, minorities, working poor and the unemployed) from unbridled greed, corruption and the desperate consequences of economic downturns for wide layers of subsistence-level labour. The 1930s Great Depression, following the 1929 Wall Street crash, culminated in the social democratic Roosevelt New Deal. In Britain, the approach was mirrored in the writings of John Maynard Keynes, the father of Keynesianism (or 'pump prime' economics based upon government funded stimulus of the economy). Both initiatives shaped the political landscape of the advanced economies for 50 years until the advent of the Reagan/Thatcher neo-liberal 'revolution' in the 1980s.

Its contemporary expression through public figures such as Canadian journalist and activist Naomi Klein and the US GND Democrats represents a nostalgic political return to the original, greater 'social equality' mission of the pre neo-liberal period. It reflects the effects of the twin capitalist crises of climate and globalisation on the wider body politic in the advanced capitalist countries. Key proposals include things such as net zero emissions by 2050, renewable energy, energy-efficient buildings, electric vehicles, job re-training for new economic development, and securing clean air and water. It also includes education, housing and health care for all.

Indeed, the ecological component of Bernie Sanders' GND presidential campaign is only one of several broad policy components designed to win the transformation mandate of disenfranchised subordinate classes and their overlapping social minority components. For example, it includes universal healthcare with 'Medicare for all' and the cancellation of student debt. In the United Kingdom, advocates for the GND – which have included non-party campaign groups as well as 'Labour for a GND' – have referred to creation of green collar jobs, public investment, reform of the banking system, greater security for pensions and savings, and environmental protection.

The Sanders' GND green jobs and emission reduction economic transitions are built upon Modern Monetary Theory (MMT) money-printing debt financing combined with greater taxation of the corporate sector. It is largely based upon State regulatory intervention and partnerships with Capital rather than post-capitalist transformation (thus still similar to the pro-capitalist European Green Deal). A core political target of the Sanders' campaign has been to win back the disenfranchised rust belt working class constituency that went to Trump over Clinton. The program is progressive insofar as it involves greater social dividends and democratic political controls over selected parts of the capitalist economy compared with neo-liberalism. But it is not transformative of that economy in a post-capitalist, socialist direction, despite Sanders' self-description as a democratic socialist.

What mass momentum there has been for a Green New Deal has now stalled. The Left social democratic electoral push depended upon the Sanders presidential nomination succeeding. That has been sidelined by the pro-capitalist centrist core of the Democratic Party, as it was in the nomination battle against Clinton in 2016. The centrists achieved this in no small part because the hoped-for activation of the 'left behind' and disengaged 45% non-voting bloc, particularly young people, failed to sufficiently materialise for Sanders. This is the second of consecutive failed nomination elections for Sanders. It comes on the back of eight years of Obama centrism and its abysmal returns on broad expectations for ecological and social transformations.

The strategy of Democratic centrism is to win the middle strata ground to defeat Trump in November. The prospects along the way have been 50/50 at best. But there is little prospect of political support for the GND worth mentioning if the pro-capitalist centrist Democrats are to succeed. Of all 20 presidential candidates the presumptive nominee, Biden, has been the *only* one to explicitly reject both the GND and social reform measures like universal 'Medicare for all'. Even if these continue to get some lukewarm endorsement to keep the Sanders activists inside the Democratic Party campaign it will only be a tactical move to eventual betrayal. The pro-capitalist policy agenda of the Democratic Party reflects that it is a political wing of a core part of the American ruling class. The most progressive candidate after Sanders, Elizabeth Warren, made it explicitly clear in the course of the campaign that she was 'pro-capitalist'. It is reflected in the Democratic Party donor base, its elected political representatives across Congress and State governments, and the weight of a supporter base among affluent working middle strata. In any event, the pandemic and its breathtaking economic consequences are overshadowing everything, especially current electoral outcomes.

The momentum for a GND has been internationalised to some extent. Its language and policy content have been included by a range of Left social democratic currents across the advanced economies. Its fate elsewhere, however, has generally been the same as the USA. Corbyn's internally conflicted left social democratic Labour Party recently crashed to a heavy electoral defeat in the UK. With it went the immediate prospect of their Green New Deal. The British Labour Party passed two motions backing a GND in late 2019, one calling for net-zero emissions by 2030, the other for establishment of a National Climate Service. Yet the GND seldom featured in its election campaign. Any public attention tended to be on the 'Green Industrial Revolution', an initiative that reinforces privatised ownership structures within the economy rather than broaching the idea of radical transformation. The Australian Greens have a GND as part of their policy agenda, but it has little public awareness or momentum beyond their own membership. The capitalist-centred European Green Deal of the EU is yet to play out in the altered budgetary landscape of the current crisis and there is resistance to many of its key provisions.

Critical political support on the part of the Left as a whole should be forthcoming if Social Democratic versions of the GND have actual and imminent prospect of electoral victory and implementation. On the brink of a choice, we need to choose the immediately achievable step in the right direction. There is no scope here for Ultra-Left sloganeering from the sidelines. While limited, its policies are nonetheless progressive enough to potentially shift the reactionary momentum on climate and class issues. There is no greater political priority and obligation in the first instance than to change the current trajectory of the climate and ecological emergency, and to contest the growth of reactionary politics. But that's a two-way obligation. It also applies to *Left* Social Democrats and how a climate action transformation agenda should now be pursued, at least if their priority political commitment is to meet the challenge of the convergent crisis and its existential impact on the global majority.

The GND electoral strategy has stalled if not failed in the immediate term at least. It's unclear what left social democracy is now advocating strategically across the advanced economies. If anything, more of the same electoral strategy with the clock ticking now demands critical re-configuration of such a minimalist GND strategy, one that explicitly combines it with a transformational nationalisation of class and corporate ownership and power. Comprehensive transformational change must be on the immediate contemporary agenda. Arguing for another social democratically framed electoral push in four more years is in itself politically hollow. The convergent ecological and economic emergency compels a critical reassessment and realignment of the approach of the broader Left movement, including on the part of left social democracy itself.

WHAT NEXT?

Importantly, the convergence and fracturing of the cross-class hegemony of capitalist rule is also being reflected in the diverse approaches to the climate question within the ruling class. This makes it a particularly critical time for the climate movement.

It is past time to assert that we are *not* in fact 'all in this together'. Those who own and control capital are culpable, not the relatively powerless global majority who need to sell their labour in order to live in a system they are unable to control. Those in power, or who dispatch power on their behalf, are quite determined to be the section of the human species that is preserved during the course of the inexorable unfolding carnage. The inequitable consequences of the coronavirus pandemic highlight the class divisions of the crisis. The poor, the destitute, the refugee, the stateless and the homeless are all especially vulnerable, not only to the virus, but because health systems expressly and explicitly have not been designed for 'them'. Victim hierarchies are recreated as Prime Ministers and Presidents lay claim to the best health care possible, while elsewhere the uninsured are turned away.

But the class division exposed by the pandemic runs substantially deeper than this. It's the frontline workers across the economy, nationally and globally, who are bearing the brunt of the pandemic. Health workers are particularly impacted. So too are the 'essential workers' in the food supply chain, from the field hands and abattoir workers through to the shelf stackers in the supermarkets to the delivery drivers. No protective lockdowns for them. 'Essential' capitalists are nowhere to be seen. Nor is there a clamour for them to show up for loan interest and rent collection.

Outside the relative affluence of the advanced economies, the working classes don't have the safety nets and health services that buffer the consequences that this long-term economic and health crisis is inflicting. Global heating will exacerbate the divisions and suffering that are already here.

So, to be clear, no matter how hard 'we' continue to push to make capitalism greener (and yes, that should continue – with qualifications centred around transition technologies and greater social justice) that doesn't mean we can make a green capitalism. Capitalist power and interests will oppose, frustrate, delay and ultimately prevent democratised and ecologically sustainable production and consumption. Prioritising the social and ecological interests and needs of the majority will be resisted by the privileged minority elite.

The political support for green *transition measures* has to be combined with a general *transformational agenda* for ecological and economic government. As with the Global Financial Crisis, re-nationalisation of certain privatised capital assets, like Public Utilities, may be achievable by social democracy in the current Covid-19 crisis. A nationalised bank that could compete with the sectoral interests of finance capital, for example, is also possible. The socialisation of bank losses during the GFC was certainly capable of leading to an actual return to publicly owned banks. That is not the same as an economy-wide socialisation of the central means of production, distribution and exchange and the guarantee of the core consumption needs of subordinate classes, nationally and globally. In any case, even limited nationalisation didn't happen post-GFC.

Likewise, promoting green entrepreneurship in the energy and food production sectors and replacing the fossil-based energy industries fuelling capitalist growth is well underway. But challenging or even reforming capitalist ownership and power over the means of production and consumption, stopping expansion of capitalist growth for private profit, and ending/ reversing the rate of exploitation and destruction of labour and the natural world, remain barely visible on the horizon. And the change across key economic sectors is hardly deep or rapid enough.

It is also, tragically, too late to avert catastrophic change even by overthrowing capitalist power, globally or even solely within a significant nation-state. 'Out of time' means exactly that, as does 'emergency', 'crises' and 'catastrophe'. Sooner or (just a little bit) later, this 'Century of Extinction' will also be history, along with millions of other 'culpable' humans (on the wrong side of our fortressed first world walls) due to climate-refracted war, starvation and disease, as well as more 'other species' than we can count.

Much as this may sound as a case for 'all hope is lost', this is not the case we are arguing. To draw a parallel with the growth of carbon emissions, the greater the delays to take action the greater the scale of transition and the costs will be. The continuing failure to green capitalism has elevated the urgency and scale of capitalist transformation beyond the reform parameters our movement has mostly advocated.

There is transformational political opportunity provided in and by the dynamic character of capitalism itself. Grim though the climate emergency is, and as omnipotent as capitalist power and control appears, there are still cracks in the edifice of global capitalism through which political opportunity and transformational hope continue to shine.

Capitalism in fact provides the structural contradictions and structural antecedents for resolving the systemic crises it has created. False hope based upon a political quicksand of trying to reconcile irreconcilable class interests is a guarantee of despair, sooner or later. Real hope is born out of necessity for a change of course, regardless of the obstacles in front of us. That's the firm ground of hope that real rebellions are built upon.

6

GREEN GLOOM, BUSTED BOOM, BARBAROUS DOOM: WHAT'S LEFT?

The current pandemic health and economic crisis crystallises the structural convergence of capitalist ecological and economic crises. This convergence will continue to deepen. It's not an outlier to 'business as usual'. It's a *political crossroads* to an intersecting emergency that cannot be undone. It is the course ahead that is in question.

The politics of the current crisis is yet to fully unfold. Emergency measures in relation to health and the quarantining of economic collapse are dominating immediate attention nationally and globally. It's difficult to imagine that the scope and scale of this economic crisis will be anything other than as historically profound as the Great Depression 90 years ago, or worse.

The ecological crisis remains profound and unprecedented. A return to capitalist growth, even a greener version, will exacerbate climate change and ecological catastrophe exponentially. Moreover, billions of people outside the affluent borders of the advanced economies are not going to pass quietly into the night as our outsized national and per capita carbon footprints continue to stride the global stage.

The political challenge of a depression-level economic crisis is not one our ecological movement has confronted. The GFC of 2008 was no where near as devastating as the present condition. For the anti-capitalist Left, our historical reference points are from an era of capitalist development and politics none of us have lived through. For some of us, at least our parents' generation shared the experiential realities of austerity, fascism and war. Like their experience, the politics that lie ahead will undoubtedly be fierce. The social divisions and difficult choices already before us will become even more stark and uncomfortable.

Ultra-Right 'solutions' to the crisis are several organisational steps ahead. Fascism prevailed in depression-era Germany against a discredited and dysfunctional State apparatus and ruling class. More important to note, it also prevailed against the largest left social democratic political party the advanced economies had ever seen. If the current economic crisis and political trajectory plays out along the lines of pre-existing scripts, then it will be a grim story indeed. For us in the broad left, we are already in the fight of our lives whether we fully realise it or not. We need to rise to the challenge of the times we are in.

In this and the concluding chapter we outline the challenge of reframing agendas for responding to this unprecedented convergent crisis. This cannot be done isolated from the conflicting class interests embodied within existing agendas. Moreover, we need to take account of the class interests within the political struggles before us and whose interests are going to prevail. We argue that a re-shaped eco-socialist strategic political direction is required. More importantly we argue that eco-socialist transformation is both possible and, perhaps for the first time, historically within reach.

WHAT'S LEFT?

Despite the climate emergency leading into the current crisis, there is a need to acknowledge the relative lack of urgency in the global political culture to avert the worst, especially given the scale of the challenge and the clarity of the science. On the plus side, the emergence of School Strikers for Climate, the recent global climate strikes, the momentum for a Green New Deal, and the growth of Extinction Rebellion have shown encouraging promise. However, the anticipated emergence of a decisive progressive movement in the advanced economies to respond to the ecological crisis has remained substantially and persistently out of reach. So, too, has the integration of subordinate working and middle strata classes into a shared struggle for progressive change and transformation.

Historically inherited forms of transformational politics – Left Social Democratic, Revolutionary Socialist, Anarchist, and their respective Green configurations – are still largely marginalised voices at best. Even in regards the best-case versions of transformational momentum within left social democracy, we see that the Sanders/GND nomination has fallen well short in gaining support for its quite limited agenda, as has the Corbyn wing of British Labour. In the immediate term, Centrism retains a firm grip on organised social democracy on both sides of the Atlantic, even as its rusted-on support within the working class is crumbling. This is certainly the case in Australia. In continental Europe, political fractures toward the left of social democracy are more pronounced, especially in France. But the same shift from centrism also applies to the Ultra-Right, more so even. On the whole, our collective Left voices remain disjointed, chaotic, insular and unfocussed. This chapter seeks to address some of these dilemmas.

Eventually the structural insanity of endless growth on an ecologically finite planet, endless consumption by materially saturated and indebted consumers, and exponential disparities in created wealth and power, can only implode. But what will be left? Trapped in a vortex of spiralling ecological, economic, and consequent geo-political crises, is a global majority perched at the precipice of descent into increasingly marginal material conditions of life. This looming future currently remains unimaginable within the still relatively affluent matrix of the advanced economies. In time, the contingent residual affluence for our subordinate class majorities too will pass. The current economic crisis may well be a watershed moment for transition into a pre-cipitous decline in material well-being.

There is an underlying despair, masquerading as states of denial, woven through the fabric of everyday political culture in the advanced economies. Whether exemplified as a rusted-on blinkered determination to cling to unsus-tainable prosperity, or dystopian resignation to its consequences, indisputable and confronting environmental facts are crucifying the minds of class-divided communities and activists alike.

A central point from preceding chapters is that the core class structure of capitalist interests and power, which foremost includes private ownership of the means of production and appropriation of the social surplus, must be overcome first before alternative post-capitalist, ecologically sustainable imaginings can be democratically fashioned. Otherwise, even the most sen-sible, science-informed and progressive 'blueprint' just becomes utopian dreaming. We know what needs to be done and how to get there from the point of view of technology, public policy and the marshalling of material and human resources. The 'why not' and 'why haven't we already done so' questions, however, return us time and again to the fundamental stumbling block – the class-based hegemony and power of contemporary global capitalism.

A reconfiguration of the strategic focus for the progressive, anti-capitalist and ecological Left must necessarily flow from these central considerations. Historically, crises in a mode of production are opportunities for political transformation, as seen in the transformations from feudalism to capitalism. Such opportunities are also present today with respect to the capitalist mode of production, particularly given that it is hard-wired to a core expansionary imperative. Talk of political openings sounds hopeful, and still can be, but these crisis transformations are not guaranteed to be positive and progressive, as 1930s European fascism and World War II remind us.

But first let's reconsider the general political picture at the moment. There is an overdue need to acknowledge that climate change has long ceased to be a fringe green, socialist, or Left issue. 'Free market fundamentalists' are far ahead of us on the political curve of the climate change debate. Global warming deniers are already morphing into geo-engineering enthusiasts. Fossil fuel corporations are diversifying into 'transition fuels' and renewables. Marketised carbon trading schemes are sprouting alongside direct action/public relations carbon sequestration offsets. 'Green' entrepreneurs, large and small, inhabit the political centre of the Green movement – both as financial backers and policy framers. Now most governments of the largest capitalist economies (and biggest aggregate carbon polluters) are 'leading' the way to moderate their climate impact whilst simultaneously expanding the very capitalist growth that will ultimately overwhelm any initiatives foreshadowed. This is all happening simultaneous to the vociferous defence of coal mining, fracking, deep-sea oil extraction and the fossil-fuel industries generally.

The overall green movement is collectively failing to adapt to the reality of the actual reciprocal structural convergences of the climate crisis within the CMP, economically and geo-politically. Instead our movement and supporters are generally latching on to any one of a multitude of capitalist climate change responses and 'solutions' as somehow 'progressive' signals that political and technological salvation remains possible within a greener capitalist framework.

One can legitimately ask whether this is a serious and deepening *delusion of hope* for what could be, or a reflection of powerless *despair and resignation* to what increasingly appears as inevitable. Each of these is bounded by similar considerations, namely, many people appear to be blinkered about:

- the inherent expansionary growth requirements and character of capitalism,

- the resilience of its accompanying structurally unequal class power structure,

- the geo-political crisis dynamics that will be unleashed as a consequence of structural growth and ecological crises, and

- the scale of the transformational dynamic that is required for retaining a liveable post-capitalist world any of us may wish to embrace.

The political economy and ecology constraints on the way forward foreshadow a bleak future for what is unfolding and concretely possible. A barbarous course of increasingly fortressed advanced nations and privileged internal enclaves versus excluded others (such as desperate climate refugees) looms large, with only the pace and form of that unfolding doom appearing to be in question. It is a challenging political pill to swallow if the presumptions and assumptions above constitute the sole line of argument. In this we see yet again the same logic as 'there is no alternative' in response to 'why capitalism' arguments. It just entrenches some form of existing capitalism as the only path ahead.

Various commentators have lately been arguing in support of the need to frame a positive hope message to successfully reach and secure an audience for mobilisation. We take the point, but the current balance of historical and structural forces is nonetheless ominous. To not frame it in these sober terms risks adding to the very delusions of the prevailing affluent political culture most of us are battling. In our view, hope does not lie in idealistic shout-outs that things for the better can happen if we unite as one (which assumes that interests are commonly shared) and if we support all policy options on the table (which includes eco-capitalist market solutions).

Our argument is that hope is grounded in the material contradictions of capitalism itself, which indicate structural fault lines that can be exploited in challenging the fundamental class dynamics underpinning global heating.

Yes, as a movement we 'need' to have hope, but this must be concrete and strategic. It is more to the point that we *can* and therefore *should* have hope. There are a number of structural crises described below that are converging within the current Covid-19 economic crisis. On the immediate horizon they threaten to further fracture the cross-class hegemony of Capital. This looming convergence represents an opportunity.

Unless the climate action movement coalesces into a political challenge to capitalist power, the recent political energy unleashed by the scale of the crisis (such as the Student Strikes For Climate, and Extinction Rebellion) is doomed to go the way of the Occupy Movement response to the GFC. That conjunctural moment and failed political response is instructive at several levels:

- the globalisation of capital, and this is most important, also globalises the structural dynamics of cyclical crises and the collapse in the accumulation process;

- the politically vulnerable and historically contingent character of capitalism as a mode of production and consumption is laid bare during such crises;

- the role of the Capitalist State, institutionally and regardless of temporary political occupants, becomes transparent – which is to preserve the capitalist order and resolve the crisis conditions to re-start accumulation;

- the structure of capitalist class power is contingent upon the sufficient active consent of subordinate classes. The extent of that hegemonic consent rests upon the capacity of the politically dominant sectors of capital to secure the immediate material interests of subordinate class sectors; and

- conversely, the incapacity to secure the immediate interests of numerically superior subordinate classes creates the political conditions for the transformation of social and political power.

The failure to seize the political momentum created by the Occupy movement revealed the lack of preparedness of the anti-capitalist movement and the 'transformative' political bankruptcy of Social Democracy in the advanced economies. Unless we change course, a similar destiny will apply to the current crisis, as it has applied to previous periods of crisis of accumulation and of heightened social and economic inequality.

The old saying, 'History repeats: first as tragedy and second as farce' looms large. Now that another GFC equivalent – except much deeper and much worse – economic and political crisis of Capital is upon us history threatens to repeat. We should already be well past politically repeating that GFC-Occupy tragedy. A continuing social democratic 'politics as usual' response to this convergent capitalist economic and ecological crisis would be to slide from tragedy into historical farce. We can't afford to let this happen, for if it does it is not just economic wellbeing that is at stake. At stake are the shared existing ecological conditions of the planet necessary for our continuing societies and other still surviving species.

A TRANSFORMATIONAL MOMENT

If there is to be hope it will be created by an increasingly disadvantaged and disenfranchised global majority forced to democratically assert its own – 'our' – collective self-interest. This will only eventuate amid collapsing ecological and economic conditions that are not of our choosing. People will react by 'taking to the streets', not because they want to, or feel morally compelled to, but because they have to. Moreover, these deteriorating conditions will be fraught with perils associated with fragmenting, constrained and contested political choices (witness Lebanon and many other countries that already are under considerable political and economic strain). But within this confronting vortex of deepening despair lies the opportunity to create a new course.

The climate action movement needs to be organically merged with an anti-capitalist movement that prioritises the ecological catastrophe. The wide layers of global working classes are the only source of majority democratic social forces available to challenge capitalist power and avoid the ecological dead-end of so-called green capitalism. Without political convergence and mobilisation with the core material insecurity, necessities and aspirations of subordinate classes the climate action movement will basically be left with cap-in-hand petitions to the halls of power, being a cheer squad for 'green' entrepreneurs, and/or street theatrics.

It is precisely the subordinate class layers globally that will be most impacted in the current crisis and as climate change deepens. The increasing crisis impacts of climate change on the material conditions of subordinate classes will be coinciding with deepening accumulation crises and re-structuring strategies to extend growth. It would be a grave mistake to suggest that the resultant impacts on subordinate classes will be met with adequate levels of transition support from Capital via the State (left social democratic GND or not) or relative political indifference on the part of those affected. On the latter, one only has to look at the dramatic shift to the Ultra-Right in the US, Britain and various parts of the EU and the current (armed no less) anti-lockdown demonstrations in the United States.

The contours of what that integrated transition plan should look like can only be broadly charted. What is clear is that building broad-based anti-capitalist support for a democratic climate action transition is a fundamentally *economic* question that requires considerable clarity about the transition to galvanise support.

A central question will be that of *growth*. Contesting the capitalist growth mantra with a 'no growth' agenda will not change the political terrain, and certainly not even get a hearing from the working-class layers we need globally. Nor would it be politically sustainable in every sense of the term. The terms of how and in what form measures to redress declining material well-being of subordinate classes within the advanced economies, and unfulfilled immediate material well-being for our counterparts outside the advanced economies, need to be reconciled. Moreover, global inequality between ruling classes and subordinate classes is increasing not decreasing.

A political course to end disparities within and between classes on an ecologically finite planet needs to be central to our endgame to save and construct *What's Left*. There is no plausible democratic mandate that can be fashioned that fails to realise the core immediate material requirements and aspirations of subordinate classes, nationally and globally. All of this has been heightened because of the coronavirus pandemic. Millions within the advantaged countries have suffered dramatic financial and employment-related losses, not to mention threats to pensions and livelihood beyond retirement. Core social needs have been reaffirmed – housing, health, education, childcare, parks and beaches – in ways that have transcended consumerist commercialisation. For many, the meeting of these needs has become cherished in ways just taken for granted before the pandemic. Shopping for shopping's sake has been shown not to be essential.

This has also raised expectations that the State can and *should* guarantee the basics for everyone. After all, it is 'the people's money' being spent on the people. By contrast, there is much public critique and YouTube mockery directed at the billionaires who queue up for no-strings-attached State handouts. Many are not backward in coming forward deriding the bailing out of the super-rich, who pay little if any tax, and yet who demand tax-payer assistance for their businesses. The *affective* dimension of Covid-19 politics cannot be discounted as a powerful motivating force for substantial change.

In this conjunctural context, the required shared conclusion amongst political activists that 'business and politics as usual can no longer be accommodated' needs to nevertheless go beyond the political, the subjective and the pejorative. It needs to reflect that structural transformation will be objectively required in the unfolding conditions of ecological catastrophe. The preceding discussion of what these threads and directions entail contributes to a more focussed approach to the stark choices in front of us.

Firestorms, droughts, floods, superstorms, heatwaves, pandemics, mass migration – all increasing in frequency and intensity – are perils that will demand response whether one 'believes' in the encompassing realities of climate change or not. Yes, mass migration is also a peril that needs confronting, as is genocide – particularly when you are on the criminalised end of fleeing or dying. These are multiple issues, an explosion of wicked problems, that confront us here and now and well into the future.

As in war, the peril of what is in front of you is there in its immediacy as a peril irrespective of the origins, motivations and miscalculations that brought it to your door. The convergent economic feedbacks, and political ramifications, mean that the question of 'What next?' or 'What is to be done?' is paramount. However, it is paramount by virtue of the objective circumstances that, regardless of hope or doom, will force a course of response to solve this convergent crisis. There is no escaping the problem. There are just confronting choices to be made in how we respond.

A TRANSFORMATIONAL CONVERGENCE

Globalised capitalism in its current form must and will change, dramatically. It cannot escape the growing climate and ecological crisis impacting upon expansion imperatives (including the limits imposed by ecology). Nor can it escape the expansionary growth imperatives generating this crisis (the pressures to accumulate). Internal systemic contradictions (especially exponential debt and underutilised capacity to produce commodities that cannot be sold) are evolved structural dilemmas stemming from the necessity to sustain accumulation. They are increasingly morphing into obstacles that constrain if not imperil these very growth and accumulation imperatives.

Also not avoidable is the impact of disintegrating cross-class hegemonic consent to diminishing material well-being emanating from these asymmetrical systemic interests being realised. The reconstruction challenges, job losses and business collapses of recent economic and ecological disasters demand political solutions. There are not many structural options extant within this historical conjuncture that can politically shape the direction this dramatic change will take. More to the point, those in power (economically and politically) will be taking substantial and substantively new steps to maintain and consolidate their interests. Failure to mobilise in ways that reflects our collective majority interests will mean that directional change in the interests of the minority in power will be imposed upon us.

Divergent class interests are at the heart of the present political fork in the road. There are also important *intra-class* divisions that influence potential avenues within the broad capitalist pathway. There is an almighty tussle unfolding between the two core wings of the ruling class in the advanced economies: the globalist, greener capitalism wing and the radical conservative, Ultra-Right nationalist wings. These internal dynamics within the context of the EU threaten to continue ripping the capitalist union apart (beyond Brexit). They are also reflected in re-configurations in the geo-political shape of the post-war world of the advanced economies in relation to NATO in the Atlantic region, the UN globally, and the realignment of multi-lateral trading blocs. It is mirrored as well globally across the less advanced capitalist States (e.g. India, Brazil). There is also a third competitor, which is the authoritarian State Capitalist form exemplified by China.

These dynamics within the global ruling class cannot be ignored by our movements, not only for their geo-political importance but for the political vortex they represent to suck our movement into political agendas against our interests. The extent to which our movement is being drawn into market-based carbon emissions trading schemes is a case in point. So too is the push to redraw multilateral trade deals because our national working class is being 'disadvantaged with lost jobs', or nationalist posturing against undemocratic European or global institutions, or border security to contain unfettered

migration from the desperate conditions of failed States and ecological catastrophe. The push for a national and international trade and 'cold' war against
China by the Trump administration looms large. More than a few progressives
are being drawn into an anti-China agenda as a priority political focus.

Consider this. We are currently in emergency conditions of capitalist government with the Covid-19 pandemic. The emergency form of capitalist government is destined to become the norm rather than the exception as rolling
climate crises and their refracted and intersecting economic and geo-political
impacts unfold. In whatever form, and regardless of consent, emergency government entails a degree of authoritarianism by default.

The forms of system-wide control required to identify, coordinate, regulate, allocate and complete eco-emergency crisis response and/or transition
requirements is beyond the capacity of national governments to democratically direct within a privatised means of production, distribution and
exchange economy. The only historically indicative parallels here are in sit-
uations of war, the so-called 'command economy' in which production is put
at the behest of the war-time effort as ordered by State decree (which finds its
contemporary parallels in admonitions for businesses to convert production
from, say, alcohol to hand sanitiser). But these are examples of 'temporary
emergencies' not as permanently transformed new states of political and
social structural re-organisation.

Under present circumstances, Chinese Capitalist State-type control may be
more relevant, and troubling, than may first appear. It signals the command,
more authoritarian state form that an integrated eco-capitalist response
would invariably take in the context of escalating crises. During the GFC, for
example, in the U.S. financial State apparatus, some 70% of the top decision-
makers who crafted the public money printing/debt-purchase solution to save
Wall Street were Goldman-Sacks executives. 'Socialisation' has many strange
bedfellows in times of accumulation crisis. The central point is that the
relative autonomy of the capitalist state (i.e. as an indispensable functional
weight within the structure of contemporary capitalism) that formed through
the twentieth century will have to be more directly merged politically with the
central corporate forms of globalised capitalist power in order for a global
response to the ecological crisis to emerge. The fiction of the separation of
economy and polity may well be superseded by the exigencies of the moment.
But this too will require the forging of renewed hegemonic consent on the
part of the ruling class.

Whether it's the globalist or the ultra-nationalist wing that prevails within and across the national boundaries of the ruling class is not a trivial concern. Nevertheless, the greater authoritarian political form of liberal capitalist government in the advanced economies will directly reflect the competitive struggle against the increasing strength of Chinese and emerging State Capitalisms and the need to rise to the coordinated, regulatory and strategic expansionary advantages they embody. Escalating ecological catastrophe within a capitalist economy will demand it.

A greener version of existing global capitalism cannot resolve the internal contradictions of the CMP and the ecological catastrophe it has unleashed. Transition to a greater coordinated and regulatory eco-capitalist State has possibility as a structural response but not in the political form some advocates may envisage (e.g., GNDs). Class interests dictate that power and control reflect private minority interests within the context of enhanced global competition. The greater fusion of State and Capital can never end in ecological and economic liberation. At best, the extremes of environmental degradation and carbon emission can be moderated in particular manifestations (for example, through pollution controls and a shift toward alternative renewable energy sources), but the trajectory of increased commodification, extraction of resources and externalisation of harm remains central to the growth imperative.

This is the dilemma that the GND advocates would immediately have confronted had they succeeded in winning government. Greater regulation and taxation of corporate power in itself would not have been (and will not be) an end in itself to resolve the ecological and economic crisis. Intervention would have to proceed to either expropriation of corporate ownership and power in the democratic interests of the subordinate class majority or collapse into authoritarian capitulation to national and transnational corporate power and its expansionary interests in competition with other authoritarian State capitalisms.

But there are wider considerations and possibilities in this period of transformational convergence. In the transition from feudalism to capitalism structural antecedents had accumulated which both exacerbated and posed transformational solutions to the contradictions in class structure, state power, technological and capital formation inhibiting further expansion.

Transformations into a new mode of production only occur historically when the contradictory forces and relations of production of the old mode reach a level of irreconcilable impasse. At that juncture a dramatic level of political and social structural transformation is required to address and resolve the impediments of the existing old order. This forms the building blocks for transition towards a new mode of production that is better equipped, more efficient, and effective, and more productive for the technological and resource capacity bound up with/inherited from within the old mode of production.

As we have previously outlined, it is necessary to distinguish the technological aspects of a mode of production from the form of economic relations within which the technology is embedded. At the same time, it is not possible to separate or disentangle technological development from those economic class relations. The old technologies of petty commodity production, the feudal order and the Absolutist State were constraints on capitalist expansion and the interests of the growing propertied classes. The technological development of the productive forces of the industrial revolution were intimately bound up with and reciprocally unleashed with this capitalist expansion.

The industrial revolution could not have been a socialist rather than a capitalist one. The social class forces for socialism were not present within the transition from feudalism to bypass the capitalist phase of generalised commodity production and the class of wage labourers it engendered. These same structural constraints underpinned the trajectory towards Stalinisation in the post-capitalist transitions that were attempted in the less developed social formations of Russia and China.

These countries had not already developed into full-blown CMPs but were economically, politically and culturally still substantially rooted in feudalism. Capitalist relations of production were comparatively embryonic to that of the advanced economies. The peasantry was the dominant social class numerically. Moreover, Russia and China still carried Absolutist State forms from pre-capitalist historical periods. Therefore, the democratisation of the State form was also embryonic. Authoritarian State forms under the emergency conditions of revolutionary transformation, especially in the context of direct threats from more economically and militarily advanced capitalist economies, became structurally consolidated, and reactionary. This prepared the ground for the fundamentally oppressive regimes that would emerge after the initial regime change.

But that was then, and this is now.

STRUCTURAL ANTECEDENTS FOR TRANSFORMATION

Historically, we are in a comparable period of transformational convergence from one mode of production to another, as with from feudalism to capitalism.

Capitalism is an authoritarian form of social production and appropriation of the social surplus. As discussed previously, the history of class societies is a history of the expropriation of land and labour, with the ruling classes in each mode of production exercising power and hegemony in ways unique to that era. Liberal Capitalism is not a State organised form of authoritarian production, as with Stalinism or Chinese State Capitalism, but it is authoritarian, nonetheless. It rests upon private ownership and control of socially produced capital (that is, we do the work, they get the profit). It is legitimated institutionally in the form of a 'right' and 'freedom' to have private property and to profit from it, and that right and freedom is also underpinned by continuing majority consent across subordinate classes. But that doesn't alter its essential authoritarian character.

All three forms of contemporary authoritarian social production have distinct features. In the case of Stalinism, the State command economy imposed a *bureaucratic collective* form of ownership/control over subordinate classes in the social production process. Under Liberal Capitalism this authoritarian control is exercised by the individual *owners of capital* or collectively as a class of stockholders via a Board and senior management operating on their behalf. Under Chinese State Capitalism, the two authoritarian forms of social production are combined. The State *command economy is fused with the authoritarian private ownership* of the central levers of the economy.

All three forms of authoritarian social production sit (nationally and globally) at the apex of social forces of production that are structurally developing in an increasingly socialised direction since the initial transformation into the capitalist mode of generalised commodity production. We explore these developments further below. At the same time, none of these authoritarian modes institutionally reflect the structural democratisation of the increasingly socialised character of these forces. They are, in fact, increasingly at odds with their further development. The privatised control and appropriation of socialised production is imposed from above. It doesn't flow from below. What this means is that the collective interests and rationale of the direct producers is constrained or undermined. Specifically, what, how and to what purpose social production is carried out, and how it's distributed, remain in the hands of a small elite minority.

The institutional structural democratisation of socialised production is what distinguishes a socialised mode of production from the corresponding absence of this democratisation with its authoritarian predecessors. The increasing contradictions of having an authoritarian ownership, control and appropriation structure at the apex of an increasingly socialised production process is what creates the conditions for fundamental transformation into a democratically organised socialist mode of production and distribution of the social surplus.

Let's be clear about two things. First, objectively we have reached the point where there is an increasingly universalised socialised production, and this lays the material foundations for transition to an eco-socialist mode of production. Second, it's not socialism if it's not democratic. Socialism has to be institutionally democratic at the point of production, appropriation and distribution of the social surplus and in its overall State form. This *structural democratisation* is the demarcation line, no matter how much spin gets spread about which form of authoritarian social production is or was better or which form of State is preferred to institutionally secure authoritarian production and appropriation. It's not socialism if it's Stalinism. Nor is it socialism if it's 'tax and regulate' social democratic capitalism without expropriation. Calling *something* 'democratic socialism' is a dead giveaway that there is something amiss, whether it comes from the left wing of social democracy or from the remnant vanguard of Leninist revolutionary socialism. There is no undemocratic socialism.

That doesn't mean that whatever socialist transformation occurs is fixed and finished on the question of democracy, its institutional form within social production, or in its State expression. What it does mean is that socialisation of the means of production, appropriation and distribution of the social surplus is the democratic platform of social and economic organisation *sui generis*. Decisions by working people for working people may be organised in diverse ways, but what is fundamental is that it is working people making the decisions. Power concentrated into elite or bureaucratic hands is, by contrast, authoritarian and inhibiting. Emancipation from economic and ecological exploitation must extend across economy and polity.

For the first time, we are at the historical point of social development where this eco-socialist transformation is both necessary and possible. The climate and ecological emergency demands it. Increasing national and global inequality demands it. The increasing economic crises of capitalist growth and accumulation demands it. The reconstruction challenge to overcome the devastating consequences of all these converging emergencies demands it. And the advanced economies, technologies and social classes now exist to make it possible.

There are four, broadly framed, existing structural antecedents compelling transition into a socialised mode of production.

Structural Democratisation of Class

Productive subordinate classes – consisting of outright commodity production wage labour, middle strata professionalised, administrative and managerial layers across the corporate and State sectors, and petty-bourgeois enterprises (particularly those contracted to corporate enterprises and the State) – are indispensable to the operation of capitalist economy but are likewise indispensable (in broad terms) to a socialised economy. Not indispensable are the capitalists that directly and indirectly own and control this subordinated labour.

Production, distribution, exchange and consumption – none of these could occur today without an extensive army of factory and agricultural workers, technicians, accountants, builders and labourers, finance officers, logistics analysts, communications experts and the list continues on. Supporting these workers is an equally extensive network of educators, trainers, health professionals, childcare workers and aged care staff. This list could be expanded as well. Production and distribution of the social surplus rests upon a large interconnected and diversified workforce, organised into a complex web of technical and social divisions of labour, and operating from local to global levels.

Democratisation, in this sense, is an extension of the socialisation of forces of production. That is, it stems from capitalist development and dependency on the collective worker but without the institutional democratisation that is required under socialism. It thus refers to the ever-increasing diversification and integration of labour and class layers into the overarching capitalist system/mode of production.

The second component of democratisation refers to *decision-making* processes. Socialism requires both a vertical and horizontal democratisation from enterprise through to economy as a whole through to the State itself. Whether located in the work sphere (decisions over what and how to produce, and including control in and over the labour process in one's workplace) or in the polity (that is, the right to participate and vote across local to national platforms), political power is dispersed rather than concentrated in elite hands. Some degree of State democratisation has nonetheless been part of winning the consent of the subordinate classes to liberal capitalist hegemony. The franchise has only been extended in the political sphere (not the workplace or labour market) and even then, it is highly circumscribed by when, how and for what. A ballot every three or four years for occupants of parliamentary office is little more than a delegated abdication of democratic rights.

The extension of the democratic franchise to non-propertied classes, including previously excluded population groups such as women, people of colour and Indigenous people, is part of this bigger picture of structural democratisation. This cultural development of democracy among subordinate classes is not to be underestimated historically. Just ask those who lived under the yoke of Stalinisation. It has other elements as well.

The relative proletarianisation of middle strata layers of employment (for example, teachers and university lecturers) diminishes the distance between social layers by reinforcing their 'unfreedom' within existing institutional structures. The integration of petty-bourgeois enterprises as central to the concentration and centralisation of national and transnational capitalist enterprise in production, distribution and market realisation (e.g. franchise system) likewise reflects the structural democratisation of subordinate classes that parallels the socialisation of the forces of production. The further they are incorporated into the larger capitalist conglomerates the less freedom they experience. The 'family' farmer, for instance, may formally own their farm, but this belies their competitive disadvantage and/or total dependence upon supermarket giants, banks and large agribusiness. Their labour is ostensibly their own, but key decisions over inputs, outputs and financial reward are made elsewhere.

These social processes reflect a reciprocal structural democratisation of capitalist dependency upon the integrated efforts of these class layers across individual enterprise and collectively across the national and global economy. Not democratised is the actual institutionalised ownership and control relationship to the means of production or the shared distribution of the social surplus produced by this collective labour.

Institutionalised structural democratisation of social production means that control in and over the labour process needs to reflect what is structurally already in the collective hands of the direct producers. Decisions about the conception and execution of work tasks need to be dispersed throughout the production-consumption chain, as do decisions about the social purposes and needs to be met by this labour. Moreover, it is not only direct producers but non-productive sections of the population – the young, the elderly, the infirm, the incapacitated – who collectively have a stake in how social needs are met. This can only really be achieved through the full socialisation of corporate enterprise (individually and across the collective central levers of the economy) and via the collective democratisation of the State Apparatus. How the latter is socially organised and institutionally operated should itself be subject to democratic decision-making and include approaches tailored to immediate local and national circumstance.

Socialisation of the Forces of Production

The Third Industrial Revolution (with AI – artificial intelligence, 3D printing and robotics) is transforming material commodity production. *Living labour* will be dramatically discarded from the production process over the next few decades. Who then will buy all the commodities that are capable of being produced by all the dead labour embodied within the machines and technology, and with what? One solution being canvassed is a universal basic income. Everyone gets an income, whether employed or not, at a level commensurate with core subsistence requirements. This ensures political stability by avoiding widespread destitution and importantly, for the ruling class, keeps intact the private accumulation of the surplus value embedded in commodities that needs to be realised in use value consumption.

If the latter sounds somewhat absurd – that is, we pay people out of the public purse so that they consume to the benefit of private enterprise – it's because it is. The private owners of the means of production need to have income dispersed to discarded wage labour so that a market for the ceaseless expansion of commodities produced is maintained. The under-utilised capacity of the existing means of commodity production is set to be expanded (through robotics for example) and yet potentially become even more under-utilised due to drop-off in affordable market demand. This is because the indebtedness of the consumer base in the advanced economies has reached saturation point.

A basic income offsets that to some degree. It doesn't offset the saturation point of the use value of the commodities themselves. The clamour for a TV in every room tends to diminish once you get to the toilet cubicle. And, of course, you need to be able to afford to buy or rent the TV rooms to begin with. Part of the expansion dynamic under neo-liberal globalisation has been to privatise public assets as commodified services and cheapen material commodities in order to expand profit, circumvent unaffordability, debt saturation and the saturation of material commodities. Hence, the continued commodification of what we can't live without, such as our drinking water, and accelerated obsolescence by developing 'must have' iPhones with three cameras. But this too is becoming ever saturated effective demand and/or getting too expensive, and there are limits to what else can be marketised in the way of commodified services. More frequently, these limits are now being set by ecology not economy.

The broader aspect of the increasingly socialised forces of production is that the growing *interdependence of production* (reflected in the growth of multi-national and transnational corporations, globalised supply and distribution chains, and integrated technologies) is increasingly at odds with the *private character of appropriation* of the social surplus, control of what gets produced, why it gets produced, and who benefits.

As outlined previously, capitalist growth is driven by the pursuit of profit. The provision and growth of jobs, wages, consumption, product innovation, etc., is only driven by this criterion and essentially a necessary means or by-product of that pursuit. We also outlined that the true cost of capitalised production is not fully reflected in the production costs. The 'free gifts' of nature are a case in point, although frequently transnational corporations engage in patent biopiracy to make it explicitly 'theirs' and no one else's including traditional owners and users. The direct privatising of nature is also evident in GMO technology and its application to flex crops such as maize and soybean. True costs are also hidden in the already socialised infrastructure (in public sector form) that subsidised the conditions of private production and accumulation (think here of roads in remote areas paid for by governments so that mining or forestry companies can profitably operate to extract the resource). Not forgetting, of course, the more transparent give-aways to individual capitals/corporate structures from the State Budgets (via tax breaks, for example). These all add up to 'free enterprise' in word rather than in deed.

Like with Kings and Nobles, are Capitalists really needed anymore? Yes, they own and control the economy by virtue of a legal-juridical structure of private property that underpins their private ownership, control and appropriation of socially produced capital. But they are also impediments to realising the further development of the socialised interdependence of the economy and the use value of commodity production itself. They are getting in the way of the next necessary stage of societal evolution, one that reflects the ecological constraints of a finite planet and the core social needs of the people on it.

Their very existence at the apex of the national and global economy is not just a massive deducted cost of socially produced capital to maintain their obscenely extravagant conditions of existence. Nor is it simply a case that their economically disruptive and destructive competitive struggle against each other to gain strategic advantage for personal gain has become inefficient in the use of scarce resources or meeting the market demand for core social needs. The central impediment is that the continued preservation and expansion of ecological and economic well-being of the national and global community and the planet can no longer be realised with them in charge. Their private self-interests can no longer be accommodated or legitimated as a ruling class.

We don't need to bring back the guillotine, though there may be a clamour for that measure depending upon national circumstances and the scale of repressive ruling class reaction to expropriation. A really nice house, a secure comfortable retirement income, and a thank you note for their 'trickle down' historical economic service is revolutionary transition enough.

Concentration and Centralisation of Capital

It is a core tendency within capitalism for private capital to get concentrated in ever fewer hands and centralised in ever larger corporations, as less competitive enterprises crumble and smaller enterprises get bought out and absorbed to created monopoly market conditions. Anti-monopoly provisions have been a legal-juridical feature of the Capitalist State on behalf of the general interests of Capital and smaller petty-bourgeois enterprises to counteract this concentrated corporate market power. Pursuing and enforcing anti-monopoly provisions have always been a particular political focus of the Social Democracy agenda to counteract the unbridled development of capitalist market and political power.

As we can readily see, such provisions have been marginally successful at best and mostly just window dressing to the reality and power of capitalist expansion. Microsoft versus Apple and Google, or Amazon versus Alibaba, or having a handful of mega-banks within a national economy does not alter the contours of unfettered and concentrated capitalist market power over our daily lives.

A couple of mega-corporations competing against each other in the same market space are a fading fairy-tale of the societal and consumer benefits of competitive capitalism. Their size and market weight bring economies of scale to consumer prices but at the cost of squeezing their workers, contractors and their small enterprise supply chains (such as small farmers supplying the large food retailers) to the point of extracting blood from stone.

Worldwide, business conglomerates, including market-oriented and profit-driven State corporations, dominate all sectors of the global economy – from cars to white goods, agriculture to beer, clothing to housing, banking to tourism. This concentration and centralisation of capital into mega-corporations is part of the overall socialisation of the forces of production. It also clears the path to eco-socialist transformation of the central levers of the economy these mega-corporations inhabit. It is easier to expropriate one large enterprise than a thousand small ones.

Economic Centrality of the Capitalist State

The relative autonomy of the Capitalist State reflects the evolution of its indispensable functional economic weight within contemporary national and global economy. Mega multi-national and transnational corporations notwithstanding, in the advanced economies at least, the State carries a

comparable economic weight to or exceeds that of even these largest of individual corporations. That doesn't mean the State does anything other than
operate in conjunction with those interests and those of other sectors of
Capital.

The economic centrality of the Capitalist State has been an unavoidable
development within capitalism itself. For all the huff and puff of neo-
liberalism and the Reaganite mantra that 'Government is the problem' the
economic centrality of the State has been undiminished within the last 40
years. Privatising public assets for commodified services to be delivered by
Capital hasn't altered that. It actually reflects the overall economic centrality
of the Capitalist State to secure the interests of capitalist expansion.

The core role of the Capitalist State is to expand capitalist growth and
accumulation. Of course, it has an array of functions. Some are more
directly economic (such as taxation, money supply and organising trade
deals) than others, such as manufacturing consent which we enumerated
earlier. But, whether directly or indirectly, the State exists and has expanded
to secure the continuing conditions for capitalist expansion. It remains
indispensable to transnational corporations and the global capitalist system
in its totality.

The critical economic centrality of the State is particularly evident in
conditions of economic crisis. Both with the GFC and in the current Covid-19
induced economic crisis, the question of whose State it really is gets revealed.
Bailing out billionaires, corporate tax-breaks, rescuing the worst climate
vandals like the oil and gas companies, and printing money via Reserve Banks
in a 'whatever it takes' approach by neo-liberal governments to re-start
accumulation, is a revelation of the core interests that are being secured. It
also reveals the power of conversion therapy that an accumulation crisis can
bring to the psychological afflictions of fiscal austerity, small government and
balanced budgets.

Also revealed, however, is whose State it needs to become, and how that
eco-socialist transition can get funded. We've been shown its potential and
its power to respond to immediate economic, ecological and social need
when it has to. Its economic centrality under capitalism is directly transformable under eco-socialism. Structural democratisation by, for and in the
interests of subordinate classes is all that is required. This is the necessity
and possibility of hope and opportunity that has become evident during the
present crises.

WHERE TO FROM HERE?

This chapter has laid out the objective structural parameters of potential transformational change. But, of course, transformation is not just a matter of underlying material conditions. It also requires a subjective element – the politically conscious intervention of subordinate classes in the struggle for change. This is the focus of our concluding discussions of eco-socialist transformation.

7

COMMON CAUSE: EQUALITY, ECOLOGY, RE-CONSTRUCTION

The core emphasis of this book is that there is a common cause to end ecological and social exploitation. This common cause is structurally shared across subordinate classes, nationally and globally.

A greater equality in material necessities is both necessary and unifying. It cannot happen without being intimately connected with the ecological limits and possibilities of an increasingly overwhelmed planet. The trajectory of the extinction curve will translate into cumulative destructive impacts on the very ecological conditions that have sustained our Holocene social development. Re-construction of our collective lives from ever-increasing catastrophes is and will continue to be an immediate and paramount consideration.

Time is up on the Holocene ecological conditions that have sustained us for the last 12,000 years. Capitalist development over the past 500 years has brought us to this crossroads. Historically, its time is also up. We have outlined why and how unsustainable growth necessarily has to come to an end. Structural contradictions and antecedents reflect and create the *necessity* and *possibility* for eco-socialist transformation. But the case for eco-socialist transformation still needs to be made and won.

The climate endgame is upon us. But transformations from one mode of production to another don't just happen, as if they are solely a subconscious causality of historical forces beyond our comprehension or control. An almighty political push is still required.

In this concluding chapter we discuss pathways to that end. A re-framing of capitalist growth is required. Also required is a broad process of expropriation of the central levers of the economy that structurally encapsulate and drive that growth. Socialisation of the private ownership and control of these *central levers of capitalist development* is an essential component. This is a distinct demarcation from, and does not include, the private property of subordinate classes that is tied to immediate material well-being and use-value consumption. Rather, our central concern is with the commanding heights of the economy.

Fracturing consent of subordinate classes to the authoritarian hegemony of Capital in conditions of economic crisis is a necessary conjunctural component to the political dynamic for transformation. Also central is coalescing subordinate class forces around an agenda for transformation that reflects their interests and resolves the respective calamities the crisis has created. Then there is the question of having the means to pursue and implement that eco-socialist agenda.

We argue that a democratic transformational nationalisation of the central levers of the economy is the broad framework that structurally knits together these objective and politically subjective requirements for eco-socialist transformation.

AN ECO-SOCIALIST TRANSFORMATION

Eco-socialism is objectively necessary. It doesn't depend upon the success or otherwise of intellectual discourse, however illuminating or exhilarating its exposition. In other words, what we *say* at a subjective level will not in itself alter the hard realities of what is occurring at the objective level. Indeed, the discourses surrounding transformation and transition have sometimes muddied the waters and obscured what is now presently in front of us.

For example, most of our foundation documents as a broad collective movement (socialist/environmentalist) are a lot to chew over, if not impenetrable to the average Jo balancing kids, job(s), mealtimes and sleep within a 24-hour period. But eco-socialism, in practice, doesn't require jumping through intellectual hoops to discover a legacy of oracle-like socialist environmental priority that really wasn't there 150 years ago. Nor does it require re-inventing theoretical legacy, however compelling. These are all very important theoretical discussions to have, but they are not decisive. If academic or theoretical discussions were decisive then we would have solved the transparent problems of capitalist economy and ecology years ago.

Objective necessity means that there is a convergence of conditions that require a solution, and that the practical options in responding to this convergence are transparent and immediate. What we choose to do, or not do, as a broad movement in the face of the extinction curve matters. What matters more, however, is what the subordinate majorities in our national societies decide to do or not do in relation to the objective impacts of this curve. They will have to do this with or without us getting our collective agenda for eco-socialist transformation sorted politically. And they will have to respond to whatever the halls of power deliver to them as conditions for their continued subordinate existence. Frankly, we have to do the same – the crunch time is here for all of us regardless of subjective disposition.

The Primacy of Meeting Social Need

In an eco-socialised economy, production is for *social need*, both how and why it's produced, for whom and with what ecological impact. Distribution of the social surplus is according to social need. This includes ecological constraints and the needs of other species, ecosystems and biodiversity.

Socialising the central means of production, through expropriation, and equitably sharing the existing accumulated wealth of the ruling class, only goes part way to redressing widespread inequality under the CMP and the scale of ecological destruction.

At a national level, in the advanced economies, this can be significantly addressed just by expropriation and re-distribution alone. However, there will still be structurally significant inequalities between the conditions of existence of subordinate class layers in the advanced economies and the less advanced economies in particular. Global disparities in material well-being will still significantly exist when transitions toward an eco-socialist mode of production (ESMP) are sufficiently supported, possible or capable of being enacted in the advanced economies.

This brings in the thorny issue of bridging global inequality with socialist growth on a finite planet in ecological crisis. Of course, much will depend upon the inherited conditions for a transition into an ESMP. But let's presume (for purposes of discussion) that this transition occurs within the next 20 years, before the scale of ecological catastrophe and consequent economic, social, political and geo-political catastrophe transcends our more immediate imaginations.

Eco-socialist growth will be required to lift stark disparities in the conditions of existence of disadvantaged and dispossessed layers of subordinate classes globally. That has to be reconciled with the ecological crisis and the already exceeded limits of capitalist growth on a finite planet.

Eco-socialist growth rests upon a number of principles, regardless of the inherited conditions, that are fundamentally distinct from what is encapsulated in capitalist growth, including its greener/eco-capitalist form. So, what is the difference?

Capitalist growth is a fundamentally authoritarian and quantitative process: the cold, hard process of egotistical profit calculation, privately owned, directed, controlled and accumulated via an ever-expanding commodity production process and centred on exploiting the surplus value these commodities contain to make more money.

Eco-socialist growth is fundamentally a democratic and qualitative process: social production is owned, directed, produced and distributed by workers collectively according to eco-social capacity and need. Material commodity production is centred on its social use value. Expansion of production and use of resources is centred upon filling priority material and social needs.

What gets produced follows these principles and is shaped by it. The new iPhone with its three cameras (how can our lives go on without that feature?) costs about US$1,100. So, the equivalent of seven iPhones will also buy/install a clean water well in Africa (approximately US$8,000). Eco-Socialised production also includes repurposing productive capacity (like making hand sanitiser instead of alcohol or medical PPE instead of industrial safety equipment) if circumstances warrant. It also includes not producing stuff for the sake of it or trying to generate a need with planned obsolescence or clever consumer-culture marketing.

It is politically fanciful and completely unnecessary that the average conditions of existence of subordinate classes in the advanced economies have to be diminished, sacrificed or expropriated for re-distribution to lift the material well-being of dispossessed layers globally. There would not be much democratic enthusiasm or votes on offer for that proposal in the advanced economies in any case.

In Chapter 2, the six pillars of social life – water, air, food, energy, shelter and security – were emphasised because they are of principal importance to address the structural vulnerability and insecurity of subordinate classes. They have been the bedrock conditions of existence underpinning human civilisation throughout the Holocene.

Under the CMP, the six pillars are commodified/commercialised for private profit or, in the case of air, trashed with indifference in its pursuit. They are most at threat for the global majority as the climate emergency continues to unfold. They are least capable of being secured for the global majority as the CMP lurches through ecologically generated crises of accumulation and investment strikes by holders of money capital and means of production, sceptical about the prospect of profit generation and realisation. To put it differently, environmental crises will reflect and reinforce existing class divisions and governmental biases.

One only has to look at recent disaster 'recoveries', such as the Australian bushfires or the Haitian earthquake or the Puerto Rico hurricane for a simple demonstration. Natural disasters initially generate socialised responses via the State and community assistance. As soon as they revert to commercialised restoration, via insurance companies, bank finance and infrastructure re-construction they mostly become recoveries in name only. Many people don't get their pre-disaster lives back for years, if at all. Personal recovery gets privatised, jobs disappear, small businesses don't re-emerge, and communities struggle or fail altogether.

In contrast to the CMP, under an ESMP, the six pillars are *guaranteed* as a *principled condition of existence* for *everyone*. So, for example, those who already own a house would keep their house (even Bill Gates, just not the other five houses). Those who have a house partly owned by banks would get their house outright. Those without a privately-owned house would get publicly funded housing. How global housing gets provided in an ecologically compatible way is a technical eco-social needs-based issue, not a commercial issue based upon what can be privately profited from or afforded. Likewise guaranteed would be the supply of ecologically sustainable energy, water and food.

The central point with re-framing growth is that the eco-socialised re-organisation of production (through to the moment of consumption) is a democratic platform for equitable resolution within the ecological and economic constraints in front of us. That doesn't mean it's going to be an easy process without difficult choices and intense differences and debate. But it's a collective process for a resolution that will be within our hands, not an imposed authoritarian solution based upon the narrow commercial interests of a tiny minority.

Democratised Expropriation and Re-distribution

Eco-Socialist transformation can only be carried out by subordinate classes acting in their own interests and pursuing resolutions that reflect their particular material circumstances. This may sound like an obvious point. But even the material interests of a narrowly defined, commodity production-based working class cannot be universalised once the exploitation of the Capital-Wage Labour relationship is discarded. Disparities across the diversity of subordinate classes, nationally and globally, will be an inherited condition for building eco-socialism. But the more immediate and pressing political issue is finding an agreed agenda for transformation that embraces this existing diversity to begin with.

It is incumbent upon any eco-socialist movement in the advanced economies to confront the question of the relative affluence of subordinate classes. This is not simply a question in relation to subordinate classes outside the advanced economies. It goes to the heart of complexities in material disparities within the advanced economies. It particularly goes to the heart of complexities in contradictory class relations across subordinate classes.

With the current Covid-19 crisis we may well draw parallels in severity and imagery with the 1930s Depression, now and in the days that lie ahead. But 'this ain't the old days', where capitalists were capitalists and workers were workers. Capital ownership relations penetrate deep into the fabric of subordinate classes.

This penetration is well beyond that of more traditionally framed middle strata where, say, small business shopkeepers are both owners of capital and may employ some workers but also have to labour themselves. Share ownership, as individual portfolios and through pension funds, and small landlordism including holiday home rentals, are pervasive across the spectrum of subordinate classes. The ideological component alone of seeing homeownership as 'a capital investment', whether it actually is personal capital (rather than just bank loan-capital) or merely a capital asset being consumed rather than expanded, complicates traditional lines of class demarcation. It follows that the ideological and political ramifications are likewise complicated, at least if we want to win eco-socialism.

Building eco-socialism, and a movement for eco-socialist transformation to begin with, has to be democratic. Otherwise it's not socialism (as explained in the previous chapter). Talk of the 99% is one thing but *doing* the 99% is a collective group walk along a complicated road.

Expropriation of capitalist ownership and control of the central levers of the economy, as well as equitable distribution of the total social surplus, has to be carried out through institutionalised democratisation by subordinate classes themselves. That will never translate into 99% agreement. The extent of material interests within layers of middle strata that dovetail with those of the ruling class are significant enough to militate against that possibility. But the extent of opposition within middle strata to socialisation can be mitigated.

Most of the penetration of capital ownership relations, and their actual material weight in the conditions of existence of subordinate classes, is quite secondary to their core existence as dependent wage labour. But it is nonetheless significant to those who are not presumptively Capitalists. Private property matters to those counting on it to provide a buffer against economic uncertainties, additional retirement income and/or future provision for children and grandchildren.

The extent and effect of expropriation is quite capable of being quarantined and transformed in ways that do not undermine the privately held nest eggs of subordinate classes. It not only has to *appear* to be quarantined. *It actually has to be quarantined.* It needs to be clearly demarcated that private property as a whole, whether as a consumed asset or for use-value capital formation, is not being expropriated. It is only the central levers of the economy and the accumulated money capital that is being expropriated. The demarcation lines, including the leveraged transformation of private capital into nationalised social capital via the State, require built-in guarantees for subordinate classes that their material interests are not being threatened. That can only occur through the democratic engagement to construct those guarantees by subordinate class sectors themselves. Otherwise it will be a political gift to Capital and/or the Ultra-Right.

Getting 51% of subordinate classes to vote for eco-socialism may sound democratic but it isn't. Nor is it sustainable. 51% can turn into 49% in a heartbeat. Eco-Socialist transformation has to be built as a coalition of subordinate interests from below. It cannot be imposed from above through delegated authority to the political apparatus of a transformed Capitalist State, no matter how much legitimated support may be invested in that apparatus to begin with.

It is the institutionalised democratisation of the subordinate class transformation process that must be foremost and ongoing in building eco-socialism. It's difficult to say in advance for all countries and circumstances the extent to which this process will be genteel or sharper-edged or somewhere in-between. Institutionalised democratisation is intergral if it's a (democratic eco-socialist) left social democratic political party voted in with a transformational agenda to carry out the Capitalist State conversion. It is also integral if it's a (democratic eco-socialist) revolutionary vanguard party invested with a transformational agenda to dismantle the Capitalist State overnight. It is also integral to the apex form of Eco-Socialist State created when the transformational dust has settled.

How that institutionalised democratisation is to be built in practice remains an open question. Undoubtedly it will have its unique national character and dynamics. But it cannot be a separate question to the pre/post eco-socialist transformation itself. A tragic historical repeat of replacing one form of authoritarianism with another would be farce personified.

A TRANSFORMATIONAL NATIONALISATION

As argued earlier, a greener version of existing global capitalism cannot resolve the contradictions of the CMP and the ecological catastrophe it has unleashed. Neither an ESMP nor a greener capitalism is a question of just being ideologically desirable. Both are already *embryonically imminent* in grappling with the scale of the economic/ecological problems global capitalism has created but can't solve. Even an Ultra-Right, Neo-Liberal/Conservative government has to respond to the consequences of the climate and ecological catastrophe. They are already doing so now in America, Australia, Hungary and Brazil with regard to the pandemic. The consequences of climate change, including mass migration and unnatural disaster events, cannot be ignored and must be dealt with concretely.

Transition into a more eco-capitalist mode of production has possibility as a structural reform response, but not in the *democratic* political form that advocates (such as the American GND) may envisage. This shift doesn't alter the authoritarian ownership and control structure of capitalist growth interests, no matter how ecologically framed. This core relation will be amplified in the emergency conditions of increasing ecological crisis. For supporters of the GND as an eco-capitalist policy initiative, substantive democratic reform will be politically subsumed if not consumed by dominant capitalist interests.

In transitional terms, national eco-socialism is no less plausible (structurally speaking) than expecting a national eco-capitalism to emerge. Both are structural political platforms that need to reflect the foundational constraints, parameters and political constituencies within the existing nationally organised economy. Transformation can only be predicated upon these existing foundations.

Moreover, a nationally organised economy is required as a necessary pre-condition to engage internationally with other nationally organised econo-mies (of whatever persuasion or disposition, such as State Capitalism) on global economic and ecological issues and institutional governance. It is likewise required to engage with the economic weight and power differentials brought to bear by transnational corporations and global finance capital in relation to domestic enterprises and the State.

This priority focus on national economy in the first instance, and on nationalisation (as per discussion below of transformational nationalisation), may seem like an obvious point for political transformation. Yet, socialists and the ecological movement have still not quite gotten to grips with 'the national question'. The focus on internationalism and globalism in regards questions of war and peace, refugees, universal working-class rights and interests, and urgent ecological action on emissions, is part of this. But these are largely aspects of demands presented to those in power, or oppositional reactions to what authoritarian power is doing. They are not a *trans-formational agenda for government.*

By this we mean that there is a need to find a national political program that can democratically integrate the diverse *material interests* of subordinate classes, *irrespective* of whether they have been convinced of the philosophical, theoretical and political merits of socialism – eco-socialist or otherwise. The emphasis on *re-construction* reflects this focus and requirement. Lives will need to be re-built as the disasters of the ecological emergency unfold. Industries and therefore jobs and communities need to be re-shaped insofar as stranded assets and global disruptions to 'business as usual' force re-calibrations in the national economy. For example, the globalised supply chain breakdowns of the present pandemic are already forcing re-thinks in the national halls of power in relation to energy security and medical equipment and supplies. How, if, and to what extent the social needs of subordinate classes are met is a necessary part of this re-construction calculus.

Focussing on the eco-socialist transformation of the national economy does not equate to a capitulation to the ideological scourges embedded within nationalism. In the first instance, we have to win the national debate for eco-socialist transformation, and this requires winning entrenched subordinate class interests predisposed to holding on to 'the devil they know'. Their diminishing material security enhances subjective levels of insecurity before it translates into 'nothing left to lose'. This is precisely what feeds jingoism, racism, isolationism and fortressing, and their Ultra-Right populist political expressions.

Left nationalism has played no small part over the years in trapping the progressive left into easy political posturing that can quickly rebound. The disintegrating subordinate class support for the 'German Capital dominated EU' and 'the Brussels Bureaucracy' has measurably been a contributed result of the nationalist culture card played by sections of British Labour and the French Left over the years. It helped give us Brexit and nearly Le Pen. Macron didn't beat Le Pen in the 2017 presidential election. Melenchon of the eco-socialist La France Insoumise did. Firstly, by getting 20% of the presidential vote in the first round, nearly matching Le Pen. More importantly, in conjunction with other Left organisations, Melenchon managed to hold the left working-class base to not go over to Le Pen or completely abstain in the presidential run-off.

In contrast, Corbyn couldn't hold the two opposing Brexit wings of the Labour party together. Johnson made deep electoral in-roads into the heart of traditional Labour by default. In Australia, left nationalist trade unionism has fed, and been fed by, racist underpinnings relating to migration and refugees and the existential threat to 'Australian jobs' over the entire course of our

European colonisation. We recognise that the national/EU political situation in Europe is more complicated than ours in Australia, or in North America as a whole. It seems incumbent upon the European Left to necessarily combine transformational nationalisation with an inter-national socialisation of economy and State within the EU context. Otherwise, subordinate class support for progressive transformational nationalisation risks sliding into an Ultra-Right, anti-internationalist narrative. This reflects the general point that a Left nationalist political agenda contains potential pitfalls that need careful handling in the face of Ultra-Right nationalism.

Transformational Nationalisation refers to using the centrality of the economic apparatus of the State to take command of the central levers of the economy. This transformation means taking the private capital that owns and controls these levers and putting them into public hands. It will necessarily include a significant degree of outright expropriation but not entirely. Leveraged buyouts based upon tactical and strategic considerations would be a central feature as well. Certainly industry-based retirement funds would be capital guaranteed in the transformation, not expropriated.

Nationalisation has had two forms under *capitalist* development. It includes provision of 'nation building' infrastructure when private capital has not been available to create the infrastructure capitalism requires to expand (e.g. roads, railways, telecommunications, electrification, public education and skills training) and defensive nationalisation that buffers the consequences of capitalist competition, development and crises on subordinate classes, and failed enterprises in key sectors of the economy (State bank, public health, nationalised airlines and/or manufacturing industry).

Transformational nationalisation embraces and incorporates these two components. Both historical forms of nationalisation have considerable support within subordinate classes. Even 40 years of neo-liberalism, including its social democratic version, has not eliminated its residual ideological support among subordinate classes. This is reflected in public and, consequently, bi-partisan support for current public debt measures in the Covid-19 crisis. It is also reflected in opinion polls that consistently show widespread support for institutions such as public health and national public broadcasters.

There is a growing awareness of the disjuncture in the credibility of neo-liberal ideology and practice post-GFC, and the post-bailout/socialisation of private sector losses. The 'free market' is and has never been left to its own devices. Nor have diverse social interests been protected 'equally' in times of crises. This is being starkly reinforced by the current Covid-19 crisis and reflected in the reaction of governments globally.

Here in Australia our neo-liberal government has instantly converted to Keynesian-style socialists overnight and embraced (temporarily at least) MMT money printing and deficit financing in a 'whatever it takes' approach to the present crisis. Their immediate goals are to maintain the conditions of enterprise, stoke effective demand for commodified goods and services, put a tourniquet on debt defaults to banks and large landlords, and secure the basic social needs of subordinate classes sufficiently enough to stem outright rebellion.

Of course, this is not a strategic re-direction back to Capitalist State forms of nationalisation. Even defensive nationalisation barely rates mention as a strategic option. It is *crisis management* driven by structural necessity and carried out by neo-liberal political actors compelled to go down a path which is in direct conflict with the ideological fibre of their very being. But it nevertheless reflects the direction of response to this structural crisis of capitalism and the shrinking options available to Capital itself, and its State minions. For example, the doubling of very basic unemployment insurance in the current crisis, and its alignment with levels comparable to the basic job retention allowance to quarantined enterprises, already resembles the ideas embodied in proposals for a 'universal basic income'. This is essentially an idea from the minion wings of Capital itself as they wrestle with the consequences of the third industrial revolution over the next few decades. But, this too, would also be part of the platform for transition to an ESMP.

Transformational nationalisation goes a strategic step beyond its nation building and defensive dimensions, whether salvaging distressed private assets or subordinate class insecurity. It is a concrete political platform that knits together the economic mechanism to achieve eco-socialist trans-formation, buffer the consequences and insecurities for subordinate classes and communities impacted by transformation, and create the engaged political support for shaping the eco-socialist transformation itself.

Central to this transformation is nationalisation to secure the six pillars of social life, in particular the food retail giants and distribution networks, water and energy companies, and finance capital within the housing sector, including the large landlords. As environmental activists might put it, the four elements – earth, air, water and sun (energy) – constitute 'the commons' that need to be put into democratic hands. It must also include nationalisation of extractive industries to begin the *democratic* process of an economically secure, ecological sustainable transition *for* and *with* the workers and communities historically dependent upon those industries.

Needless to say, the real (often hidden) extent of public subsidisation of privatised ownership and profit, for example for all existing and future energy grids, is such that socialisation would likely be cheaper to the 'public purse', especially in the medium to longer term. More importantly, meeting the social and economic needs of the majority through democratically decided and implemented nationalisation, places socially rational outcomes over profit outcomes at the centre of technological and resource applications.

The technological evolution and revolution of ecological transition (demands and elements of which now already exist, at least embryonically) can be completely integrated with a nationalised transition into an eco-socialist mode of production, just as the *central means of production* in the CMP generally can be socialised. That's why eco-socialists can form transitional coalitions with eco-capitalists on many issues relating to the transition from emission-intensive industries.

Transformational nationalisation is the politically comprehensible and deliverable eco-socialist alternative to privatised green capitalist enterprise. In some circumstances it may constitute a public/private enterprise transition. The main point is that it's a *transitional* component to socialisation, not an end in itself.

Even more to the point, the subordinate classes we need to coalesce into a political movement for crisis response and transformation must easily be able to locate themselves as democratically engaged within a national transition program for a majority community of interests they can identify with. Socialisation requires democratic nationalisation as the first major political step. Doing democratic socialisation is infinitely more important than ideological conversion to socialism or being able to articulate its theoretical components. Supporters can read Marx and eco-socialist manifestos at their leisure.

The scope and pace of democratic nationalisation will have a considerable practical component to actual implementation. Fundamentally, it offers an integrated program for alternative government in the overall eco-socialist transformation. Nationalisation satisfies the platform requirements for political integration of subordinate class interests and security concerns. It also reflects the structural antecedents to eco-socialism already imminent within capitalist development itself. The economic centrality of the State has already been on full display in the last two global crises in accumulation. So too the hidden money tree for stressed capitalist interests has also become transparent.

The question of why this centrally organised money tree isn't available for *ecological transformation* and for securing the increasingly imperilled interests of subordinate classes is now on the national agenda unlike any period since Rooseveltian times. Surely, we can find broad left agreement for a democratic nationalisation to win the structural transformation the times demand.

BUILDING A SUBORDINATE CLASS AGENDA

It seems customary to list some prescriptive demands at the end of books on political transformation. We have none to offer, not even for our own country much less the political struggles in others. These have to flow from the coalition of social and political forces that are required as a precondition to transformation. We do, however, have some concluding observations on the directions that follow on from our discussions above.

The central question is how to move forward from where we are, to contest the authoritarian power structures in front of us. We need to win power in order to implement the economic, political and especially ecological changes required. Unless the organisations for transformation across our broad left spectrum are prepared to break out of their political silos, then finding the social class forces to contest and win power becomes a moot point. The difficulty of finding the terms for building that coalition within our fractured Left environment is not an uncommon lament. But it doesn't alter the need to rise to the challenge of the times we are in.

We noted earlier that the climate action movement needs to be organically merged with the anti-capitalist workers movement which, reciprocally, needs to reflect a priority focus on the climate and ecological emergency. The wide layers of subordinate classes globally are the only source of majority democratic social forces available to avoid the ecological dead-end of so-called 'green capitalism'.

Moreover, it is precisely the more disadvantaged working-class layers nationally and globally that will be most immediately impacted as climate change deepens. By *organically merged* we mean that the interdependent fate of subordinate classes and the ecological conditions that will sustain our collective future is now indistinguishable. The exploitation of nature, and the exploitation of us, needs to end.

To realise this we need to be guided by transition pathways that emerge from concrete analyses of systemic processes and dynamics. For us, these pathways include the following:

- The source of the climate crisis is located as being capitalism and the solution is its necessary *transformation* into an alternative, democratically owned and controlled eco-socialist mode of production;

- Transitional climate actions/demands should *fracture* rather than side-step/circumvent or reinforce conditions of capitalist wealth and power and its cross-class hegemony, particularly with regard to working class layers;

- Private capitalist market-based technological solutions to the ecological crisis should be either countered/or provisionally supported with *nationalised public ownership* and *democratically controlled* transitional solutions;

- Transitional demands for ecological transformation should counter-hegemonically integrate and guarantee the interests of the *immediate material conditions of existence* of affected working class and vulnerable/discontented middle layers nationally and in local communities;

- Anti-capitalist and transitional political demands to improve the deteriorating economic and ecological conditions of subordinate classes should be *integrated into a broader climate action transition*; and

- An integrated nationalised economic and climate action transition plan needs to be developed in conjunction with a *political transition plan for alternative national government.*

To win majority democratic support we need to be able to articulate our comprehensive alternative for government. If we are to escape the extinction curve, then our eco-socialist coalition against planetary extinction needs to formulate an escape plan for a government that our diverse support base can coalesce around.

ENDING AT THE BEGINNING

When we started our political lives the conditions for eco-socialist transformation looked a great deal more promising. The protests against the Vietnam War were in full swing. Big gains had been won by the black rights and women's movements. The modern environmental movement was emerging in response to the consequences of capitalist development. Trade union membership still numbered in the tens of millions. Members of socialist parties across the advanced economies numbered in the tens of thousands. For all the transparent flaws and horrors of Stalinisation in the post-capitalist USSR and China, a more properly socialist society looked historically possible. It just seemed a political revolution away.

For our Left generation of Boomer elders, the political circumstances of our current period look glum compared to the political heydays of the 1960s and 1970s. In the interregnum, the retreat into affluence hasn't helped. It's understandable, though perhaps somewhat less forgivable, and it has had a significant retarding influence on Left politics generally. Inter-generational affluence has afflicted all of us in the advanced economies. The price has been high, and the ecological debt is now due.

However, for all the political challenges that lie before us, and the steepness of the climb, despair is not an option. Not for the younger generations.

But nor is it objectively required. A core message of this book is that, paradoxically, the antecedent structural conditions for eco-socialist transformation have been enhanced, not diminished, over the course of our Boomer lives. The political conditions, as indicated, not so much. But that too can change in an instant.

As we sat down to write this book in January 2020, historic bushfires followed years of historic drought. The endless trickle of taunts about 'woke inner-city greenies' has dried up. The national conversation changed overnight to what to do about climate change, across the political spectrum. That is, until the coronavirus pandemic hit in March and we went from writing about the last great global economic crisis to the one we are currently living through. Now (June), in the midst of deadline week for the book, a group of white cops extinguished the life of yet another black man for a possible $20 misdemeanour. On cue, a white male president has threatened the protest movement with 'vicious dogs' and an unleashed military occupation of the streets. Books of political fiction couldn't make this stuff up.

So, in this and many other ways, we feel we are ending at the beginning of where we came in politically 50 years ago. It's difficult to say where any of this will end up politically in the months and years ahead. But no one is going to hand a better future to us. That is the challenge before us all and the clock is ticking louder by the year. Hence this book.

The extinction curve is unrelenting. In about 20 years Mother Earth is going to take her ecological response to a whole new level. If that isn't self-evident in the current ecological catastrophes, then you haven't been paying attention. The science tells us so. Our hearts and minds and instincts tell us so. We are in the midst of profound change and possibility. Where this all leads is still within some measure of our collective control.

At a broad political level, we hope this book provides a stimulus to the urgent discussions and coalitions that need to be built in order to turn objective conditions for eco-socialist transformation into the political force that is required. Central among these is democracy and how democratisation will be radically protected and extended in the course of socialisation. There is much to debate and engage with around this vital question.

In our view, no matter how dire it may look, the further the extinction curve advances the more the structural conditions exist for eco-socialist transformation to take place. It may not be in conditions that we like. Politically it could even get darker first if the Ultra-Right prevail. The Biden's, Macron's, and Merkel's are not going to save us. Neither will the next Sanders, to be frank. Not unless the agenda moves to the sort of eco-socialist *transformation* we have outlined.

Above all, Mother Earth will have her say in this post-Holocene transition. There is a climate endgame. Hubris around our human exceptionalism won't alter that. It may not be the end of all of us. But if we had to make a remnant species survival choice, we would choose koalas over the Trumps of the world.

The course of history is altered by ordinary people called upon to do extraordinary things. A singular act of defiance declaring *enough* can echo around the globe. This historical moment calls on all of us.

Real, tangible, transformational hope flows from the convergence of necessity and possibility. For all the challenges and setbacks, the battle is still in front of us. The future is still in our hands. The objective transformational conditions are in our favour. We are the majority. So, collectively, yes we can. While we, the increasingly dispossessed, live and breathe there is hope. And revolutions are built on hope.

FURTHER READING

We have drawn upon a number of books, articles, news media, and government and NGO reports in preparing this commentary. Our core influences flow from ongoing ecological and socialist movement discussions of eco-socialist issues. Contributions from 350.org, Greenpeace, Extinction Rebellion, the *Guardian* and Green Left Weekly are valuable. Additional ongoing sources include CNN, *The New York Times*, *The Washington Post*, the Australian Broadcasting Corporation and other commercial and public broadcasting outlets. We are also grateful for the continuing contributions to public debate by Bill McKibben, George Monbiot, Naomi Klein, Michael Mann, and other climate activists, scientists and commentators.

Below, we identify some of the key readings which have informed our presentation and arguments throughout the book. We have provided these on a chapter-by-chapter basis. These are indicative rather than comprehensive lists intended for those wishing to engage in further reading on these topics.

CHAPTER 1: AT DANTE'S GATE

Athanasiou, T. (1996). *Divided planet: The ecology of rich and poor*. Boston, MA: Little, Brown and Company.

Gore, A. (2006). An inconvenient truth. (Documentary film, directed by Davis Guggenheim. Lawrence Bender Productions).

Greig, A., & van der Velden, J. (2015). Earth hour approaches. *Overland*, 25 March. Retrieved from https://overland.org.au/2015/03/earth-hourapproaches/

Hamilton, C. (2010). *Requiem for a species*. Sydney: Allen & Unwin.

Hansen, J. (2009). *Storms of my grandchildren: The truth about the coming climate catastrophe and our last chance to save humanity*. New York, NY: Bloomsbury.

IPCC (Intergovernmental Panel on Climate Change). (2013). *Working group I: Contribution to the IPCC fifth assessment report climate change 2013. The physical science basis: Summary for policymakers*. Retrieved from www.ipcc.ch/pdf/assessment-report/ar5/wg1/WG1AR5_SPM_FINAL.pdf

Marx, K. (1964). *The economic and philosophic manuscripts of 1844*. New York, NY: International Publishers.

Steffen, W., Dean, A., & Rice, M. (2019). *Weather gone wild: Climate change-fuelled extreme weather in 2018*. Sydney: Climate Council of Australia.

UN Economic and Social Council. (2010). Study on the need to recognize and respect the rights of Mother Earth. *Permanent Forum on Indigenous Issues*. Ninth session. New York: UN. [E/c.19/2010/4 – pp. 1–17].

United Nations. (2015). Framework convention on climate change. Conference of the Parties, Twenty-First session, Paris, 30 November to 11 December 2015, Adoption of the Paris Agreement (includes Annex: Paris Agreement). Paris: UN.

Wallace-Wells, D. (2017). The uninhabitable earth. *New York Times Magazine*, July 10. Retrieved from http://nymag.com/daily/intelligencer/2017/07/climatechange-earth-too-hot-for-humans.html

CHAPTER 2: BEYOND THE HOLOCENE EDGE

Christoff, P. (Ed.). (2014). *Four degrees of global warming: Australia in a hot world*. London: Routledge.

Diamond, J. (2005). *Collapse: How societies choose to fail or survive*. London: Penguin Books.

Flannery, T. (2005). *The weather makers: The history and future impact of climate change*. Melbourne: The Text Publishing Company.

Fukuyama, F. (1992). *The End of history and the last man*. New York, NY: Free Press.

Intergovernmental Science-Policy Platform on Biodiversity and Ecosystem Services (IPBES). (2019). *The IPBES global assessment report on biodiversity and ecosystem services*. Paris: IPBES.

IPCC (Intergovernmental Panel on Climate Change). (2014). *Climate change 2014 synthesis report, approved summary for policymakers*. Retrieved from www.ipcc.ch/pdf/assessment-report/ar5/syr/AR5_SYR_FINAL_SPM.pdf

Lynas, M. (2007). *Six degrees: Our future on a hotter planet*. London: Harper Collins.

Manne, R. (2012). A dark victory: How vested interests defeated climate science. *The Monthly*, August, 22–29.

Rainforest Action Network. (2020). *Banking on climate change: Fossil fuel finance report 2020*. Rainforest Action Network, Banktrack, Indigenous Environment Network, Oilchange International, Reclaim Finance, Sierra Club. Retrieved from www.ran.org

Steffen, W., Broadgate, W., Deutsch, L., Gaffney, O., & Ludwig, C. (2015). The trajectory of the Anthropocene: The great acceleration. *The Anthropocene Review*. doi:10.1177/2053019614564785

United Nations Environment Programme. (2019). *Emissions gap report*. Geneva: UNEP.

Watts, N., Adger, W. N., Ayeb-Karlsson, S., Bai, Y., Byass, P., Campbell-Lendrum, D., ... Costello, A. (2017). The lancet countdown: Tracking progress on health and climate change. *The Lancet, 389*, 1151–1164.

CHAPTER 3: THE EXTINCTION CODE WITHIN THE CAPITALIST GROWTH PROTOCOL

Bakan, J. (2004). *The corporation: The pathological pursuit of profit and power*. London: Constable.

Crosby, A. (1986). *Ecological imperialism*. Cambridge: Cambridge University Press.

Foster, J. (2002). *Ecology against capitalism*. New York, NY: Monthly Review Press.

Gilding, P. (2011). *The great disruption: How the climate crisis will transform the global economy*. London: Bloomsbury Publishing.

Harvey, D. (2010). *The enigma of capital*. London: Profile Books.

Mandel, E. (1968). *Marxist economic theory*. London: Merlin Press.

Marx, K. (1954). *Capital (Vols. 1, 2, 3)*. Moscow: Progress Publishers.

Marx, K. (1973). *The grundrisse: Foundations to the critique of political economy*. New York, NY: Vintage Books.

Meadows, D., Randers, J., & Meadows, D. (2004). *Limits to Growth: The 30-year update*. White River Junction, VT: Chelsea Green.

Michalowski, R., & Kramer, R. (2006). *State–corporate crime: Wrongdoing at the intersection of business and government*. New Brunswick, NJ: Rutgers University Press.

Tombs, S., & Whyte, D. (2015). *The corporate criminal: Why corporations must be abolished*. London: Routledge.

CHAPTER 4: FRACTURING CONSENT: MINIONS, MERCENARIES, MALCONTENTS AND LES MISÉRABLES

Mandel, E. (1975). *Late capitalism*. London: Verso.

Marx, K. (1968). The 18th Brumaire of Louis Bonaparte. In Marx & Engels (Eds.), *Selected works*. New York, NY: International Publishers.

Marx, K., & Engels, F. (1968). Manifesto of the communist party. In Marx & Engels (Eds.), *Selected works*. New York, NY: International Publishers.

Miliband, R. (1973). *The state in capitalist society*. London: Quartet Books.

Piketty, T. (2014). *Capital in the twenty-first century*. Cambridge, MA: The Belknap Press of Harvard University Press.

Poulantzas, N. (1975a). *Political power and social classes*. London: New Left Books.

United Nations, Human Rights Council. (2019). *Climate change and poverty: Report of the special rapporteur on extreme poverty and human rights*. A/HRC/41/39-24 June-12 July 2019, Agenda item 3 (pp. 1–19).

van der Velden, J., & White, R. (1996). Class criminality and the politics of law and order. In R. Kuhn, & T. O'Lincoln (Eds.), *Class and class structure in Australia*. Melbourne: Longman.

White, R. (2014). Environmental insecurity and fortress mentality. *International Affairs*, *90*(4), 835–852.

Williams, R. (1977). *Marxism and literature*. Oxford: Oxford University Press.

Wright, E. O. (1985). *Classes*. London: Verso.

CHAPTER 5: REBELLING FOR A GREEN CAPITALISM IS A DEAD END

Broad. (2006). How to cool a planet (maybe). *New York Times*, June 27. Retrieved from http://www.nytimes.com/2006/06/27/science/earth/27cool.html?

Hamilton, C. (2013). *Earthmasters*. Sydney: Allen & Unwin.

Hulme, M. (2014). *Can science fix climate change? A case against climate engineering*. Cambridge: Polity Press.

Jackson, T. (2011). *Prosperity without growth? Economics for a finite planet*. London: Earthscan from Routledge.

Klein, N. (2014). *This changes everything: Capitalism vs the climate*. London: Allen Lane.

Klein, N. (2019). *On fire: The burning case for a green new deal*. London: Allen Lane.

Monbiot, G. (2006). *Heat: How to stop the planet burning*. London: Allen Lane.

O'Connor, J. (1994a). Is sustainable capitalism possible? In M. O'Connor (Ed.), *Is capitalism sustainable? Political economy and the politics of ecology*. New York, NY: The Guilford Press.

Rifkin, J. (2011). *The third industrial revolution: How lateral power is transforming energy, the economy, and the world.* Basingstoke: Palgrave Macmillan.

Stern, N. (2007). *The economics of climate change.* London: Cambridge University Press.

Whyte, D. (2020). *Ecocide: Kill the corporation before it kills us.* Manchester: Manchester University Press.

CHAPTER 6: GREEN GLOOM, BUSTED BOOM, BARBAROUS DOOM: WHAT'S LEFT?

Adler, P. (2019). *The 99% economy: How democratic socialism can overcome the crises of capitalism.* New York, NY: Oxford University Press.

Fine, B. (1984). *Democracy and the rule of law: Liberal ideals and Marxist critiques.* London: Pluto Press.

Global Scenario Group, Raskin, P., Banuri, T., Galloopun, G., Gutman, P., Hammond, A., Kates, R., & Swat, R. (2002). *Great transition: The promise and lure of the times ahead.* Boston, MA: Stockholm Environment Institute.

Kramer, R. (2020). *Carbon criminals, climate crimes.* New Brunswick, NJ: Rutgers University Press.

McKibben, B. (1989). *The end of nature.* New York, NY: Anchor Books.

Monbiot, G. (2012). We cannot wish Britain's nuclear waste away. *The Guardian*, 2 February. Retrieved from http://www.theguardian.com/environment/georgemonbiot/2012/feb/02/nuclear-waste

O'Connor, J. (1994b). Is sustainable capitalism possible? In M. O'Connor (Ed.), *Is capitalism sustainable? Political economy and the politics of ecology.* New York, NY: The Guilford Press.

Poulantzas, N. (1975b). *Classes in contemporary capitalism.* London: Verso.

Ruddiman, W. (2005). *Plows, plagues and petroleum.* Princeton, New Jersey, NJ: Princeton University Press.

Trotsky, L. (1972). *The revolution betrayed: What is the Soviet Union and where is it going?* New York, NY: Pathfinder Press.

Trotsky, L. (1980). *The history of the Russian revolution*. New York, NY: Monad Press.

CHAPTER 7: COMMON CAUSE: EQUALITY, ECOLOGY, RE-CONSTRUCTION

350. org. (2018). 'Overview' and 'principles'. Retrieved from www.350.org

Foster, J. B. (2009). *The ecological revolution: Making peace with the planet*. New York, NY: Monthly Review Press.

Global Campaign to Demand Climate Justice. (2018). 'Fight for climate justice!' Retrieved from www.demandclimatejustice.org. Accessed on March 2, 2018.

Kovel, J. (2002). *Enemy of nature: The end of capitalism or the end of the world?* New York, NY: Zed Books.

Kovel, J., & Lowy, M. (2001). Eco-socialist manifesto. *Capitalism, Nature, Socialism*. Paris. Retrieved from http://www.cnsjournal.org/manifesto.html

Lovelock, J. (2006). *The revenge of Gaia: Why the earth is fighting back-and how we can still save humanity*. London: Allen Lane.

Lowy, M. (2018). Why ecosocialism: For a red-green future. *Great Transition Initiative* Retrieved from https://greattransition.org/images/Lowy-Why-Ecosocialism.pdf. Accessed on May 11, 2020.

Meiskins-Wood, E. (1986). *The retreat from class*. London: Verso.

Spratt, D., & Sutton, P. (2008). *Climate code red*. Melbourne: Scribe Publications.

Trotsky, L. (1971). *The struggle against fascism in Germany*. New York, NY: Pathfinder Press.

Wall, D. (2010). *The rise of the green left: Inside the worldwide ecosocialist movement*. London: Pluto Press.

INDEX